Student Solutions Manual

Mathematics for Teachers
An Interactive Approach for Grades K–8

FOURTH EDITION

Thomas Sonnabend

Prepared by

Tod Shockey

BROOKS/COLE
CENGAGE Learning

Australia • Brazil • Japan • Korea • Mexico • Singapore • Spain • United Kingdom • United States

For product information and technology assistance, contact us at **Cengage Learning Customer & Sales Support, 1-800-354-9706**

For permission to use material from this text or product, submit all requests online at **www.cengage.com/permissions**
Further permissions questions can be emailed to **permissionrequest@cengage.com**

ISBN-13: 978-0-495-56170-5
ISBN-10: 0-495-56170-3

Brooks/Cole
10 Davis Drive
Belmont, CA 94002-3098
USA

Cengage Learning is a leading provider of customized learning solutions with office locations around the globe, including Singapore, the United Kingdom, Australia, Mexico, Brazil, and Japan. Locate your local office at: **www.cengage.com/international**

Cengage Learning products are represented in Canada by Nelson Education, Ltd.

For your course and learning solutions, visit **www.cengage.com/brookscole**

Purchase any of our products at your local college store or at our preferred online store **www.ichapters.com**

Printed in the United States of America
1 2 3 4 5 6 7 12 11 10 09

Mathematics for Teachers: An Interactive Approach for Grades K – 8 4th Edition
Student Solutions Manual
Thomas Sonnabend

Chapter 1

1.1 Lesson Exercises

LE1   Opener

  a) 4, 8, 18, 9, 5

  b) 5

  c) 5, 10, 20, 10, 5, 5

  d) You always end up with 5.

  e) Yes.   n, 2n, 2n + 10, $\dfrac{2n+10}{2}=n+5$, Now subtracting five yields $n+5-n=5$

LE2   Reasoning

The product of two odd numbers is always an odd number.

LE3   Reasoning

 Every time I go on a picnic, it rains.

LE4   Reasoning

 False. $3+3 \neq 3 \bullet 3$

LE5   Reasoning

  $1+3+5+7+9=5^2$

  $1+3+5+7+9+11=6^2$

  $1+3+5+7+9+11+13=7^2$

They are true.

LE6   Reasoning

 No. If the negative number is "more negative" than the positive number is positive, the answer will be

negative. For example, $1+(-9)=-8$.

LE7   Reasoning

a) Yes, the sum of any three consecutive whole numbers is divisible by 3.

b) No, the sum of any four consecutive whole numbers is not divisible by 4.

c) Yes, the sum of any five consecutive whole numbers is divisible by 5.

d) The sum of N consecutive whole numbers is divisible by N when $N =$ any odd number.

LE8   Summary

Inductive reasoning is a process of making a generalization based upon a limited number of observations.

Inductive reasoning doesn't always lead to a true generalization, but many times it does. Only one

counterexample is needed to prove a generalization false.

1.1 Homework Exercises

1.   Pick any number: n

Add 8: $n + 8$

Multiply by 2: $2(n + 8)$

Subtract 10: $2n + 16 - 10$

Subtract your original number: $2n + 6 - n = n + 6$

a) Pick 6   Pick 8

6 + 8   8 + 8

2(14)   2(16)

28 − 10   32 − 10

18 − 6   22 − 8

12   14

(6 + 6)   (8 + 6)

b) General pattern: If n is your number, the result is n + 6.

3.   Yes, you look at a limited number of specific examples to form a generalization. However, sometimes

inductive reasoning leads to a false generalization.

5.   Only one counterexample is needed to disprove a generalization.

7.   False. Mike does not drink beer, but likes to watch football on TV.

9.   Reasonable.

11. a) $123456789 \times 36 = 4444444404$

b) $123456789 \times 45 = 5555555505$

c) $123456789 \times 54 = 6666666606$
$123456789 \times 63 = 7777777707$

d) Since $123456789 \times 9 = 1111111101$, the above patterns are always some multiple of 9 times 123456789.

(36 is four times 9, 45 is five times 9, 54 is six times 9…)

13. a) On Mars her weight might be $0.38 \bullet 400 = 152$.

b) It doubles.

c) It appears every 100 lbs on earth is equivalent to 38 pounds on Mars by observing the results in the table.

d) $M = 0.38E$

15. A 5th grade teacher tries a certain lesson 3 years in a row, and each class finds it interesting. The teacher generalizes that all 5th grade classes will find it interesting.

17. a) True. Think about how multiples of 3 are situated in the whole numbers, every third one is a multiple of three.

b) True. Think about how multiples of 4 are situated in the whole numbers, every fourth one is a multiple of four.

c) True.

d) The product of any N consecutive whole numbers is divisible by N if N is odd.

19. No.

1.2 Lesson Exercises

LE1 Opener

For student.

LE2 Reasoning

Sharky stole the painting.

LE3 Reasoning

Jane did not steal the painting.

LE4   Reasoning

AB = CD and AD = BC.

LE5   Reasoning

264 is an even number.

LE6   Concept

a) If $x = 4$ and $y = x + 4$

b) then $y = 7$

LE 7   Reasoning

a) Suppose distinct lines a and b in a plane are not parallel. Therefore a and b are intersecting.

b) The first two sentences about the lines are hypotheses and the conclusion is "a and b are intersecting."

LE 8   Reasoing

All doctors finish high school.

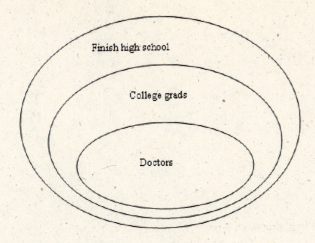

LE9   Concept

a) All doctors like roller skating.

b) The phrase after "If" and before "then" contains the two hypotheses, and the conclusion is "all doctors like

roller skating."

LE10   Reasoning

All squares have six sides.

LE11   Reasoning

a) Maija is a college graduate.

b) Melissa is not a doctor.

c)

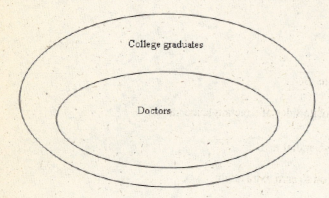

(d)

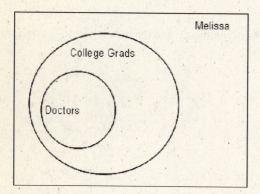

Melissa is not in the "circle" of college graduates. So Melissa cannot be a doctor.

LE12   Reasoning

(a) Since Marcus is not a doctor you could make two conclusions, he is a college student, or he is not a college student.

(b) Since Misha is a college graduate you cannot make any conclusions since the if-then statement is about college students and doctors.

LE13   Reasoning

For student.

LE14 Summary

Deductive reasoning is the process of reaching a necessary conclusion from given facts or hypotheses. When deductive reasoning is done correctly it is valid. Deductive reasoning is invalid when the conclusion does not follow from the hypotheses.

1.2 Homework Exercises

1. All rectangles are quadrilaterals.

3. a) $\overline{AB}$ is parallel to $\overline{CD}$ and $\overline{AD}$ is parallel to $\overline{BC}$.

    b) Hypothesis: $ABCD$ is a rectangle. The opposite sides of a rectangle are parallel.

    Conclusion: $\overline{AB}$ is parallel to $\overline{CD}$ and $\overline{AD}$ is parallel to $\overline{BC}$.

5. a) If you are a student, then you want the option to earn extra credit.

    b) If the figure is a rectangle, then it has four sides.

7. The hypothesis is true.

9. a) −5 is less than 10.

    b) 12 is not negative.

    c)

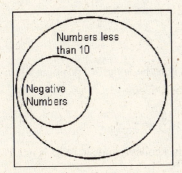

    d)

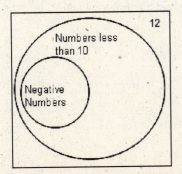

11.   a) They will get at least a B in the course.

     b) Nothing

     c) She did not get an A on the final

13.   a) Nothing

     b) That person does not live in Nevada

15.   True

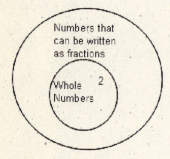

     Deduce the conclusion, 2 can be written as a fraction

17.   Yes.

19.   5, since one correct digit was lost by changing 5 to 4.

21.   a) 0, 1, 2, 6, 7, 9

     b) 8. The only digit that is dropped from the 3rd guess to make the 4th guess is 8.

     c) 5 and 3

     d) 35_ because 3 and 5 are correct digits in the wrong place. Since 8 must be in the 3rd position, fill in 3 and

     5 in the 1st and 2nd places, respectively.

     e) 358

23.   If you have your teeth cleaned twice a year, then you will lose fewer teeth.

25.   Some people do not care about their surroundings.

27.   a) All mathematics teachers love mathematics.

     b) Time flies when you are having fun.

29.   All elephants are good dancers.

     Alice is an elephant.

     Alice is a good dancer.

31.   Sandy is a female dachshund that is not white.

33. Nancy, Ma, Igor, Migraine, Lurch

35. a) 1 honest 99 crooked

b) When you pair the honest politician with any other politician, the other is crooked.

37. Select one ball from W and Y and you can determine if it's really W or Y, then work out the rest.

39. a) A 1 can be placed in row 7 column 6. A 2 can be placed in row 6 column 7.

		2		1				
	4		3		6		1	2
						3		
5	2	9	1			4	6	
4			7	9		8	5	
						2		
2		5		4	1	9		
7				3	8	1	2	
	1					5		

b)

3	5	2	9	1	7	6	4	8
9	4	8	3	5	6	7	1	2
1	6	7	8	2	4	3	9	5
5	2	9	1	8	3	4	6	7
4	3	6	7	9	2	8	5	1
8	7	1	4	6	5	2	3	9
2	8	5	6	4	1	9	7	3
7	9	4	5	3	8	1	2	6
6	1	3	2	7	9	5	8	4

41.   a) A (Dr. Howell), B (schoolmaster Wilton), C (murderer, General Scott), D (Admiral Jennings)

b) The schoolmaster does not drink alcohol, and Scott had a beer, so I deduced that Scott was not the

schoolmaster.

1.3 Lesson Exercises

LE1   Opener

Inductive reasoning.

LE2   Concept

Deduction.

LE3   Concept

Induction.

LE4   Concept

Deduction.

LE5   Reasoning

a) 4   5
$4-3=1$ $5-3=2$
$2(1)=2$ $2(2)=4$
$2-4=-2$ $4-5=-1$
$-2+6=4$ $-1+6=5$

You should always end up with the original number.

b) n
$n-3$
$2(n-3)$
$2(n-3)-n=n-6$
$n-6+6=n$

LE6   Concept

a) Ofala has not proved it since she hasn't considered all the cases.

b) Considering Betsy's work, an odd number is an even number plus one, $2n+1$, so this reasoning of adding

two odd numbers, $2n+1+2n+1=4n+2$, which is always an even number.

LE7   Summary

Inductive reasoning looks at many specific examples and forms a generalization. Deductive reasoning takes

an accepted generalization and applies it to a specific case.

LE8   Opener

a) True.

b) If a number is even, the number is divisible by 10.

c) Definitely false.

LE9   Concept

a) True.

b) If $x = 2$, then $x + 3 = 5$.

c) True.

LE10 Skill

a) Yes

b) A whole number is divisible by 10 if and only if the ones digit is 0.

LE11 Summary

"A if B" means "If B, then A."

"A only if B" means "If A, then B."

The converse of an if-then statement may or may not be true.

1.3 Homework Exercises

1.   Inductive.

3.   Induction.

5.   Induction.

7.   Induction

9.   a) It will increase by 4.

b) It may not work for some number you didn't try.

c) $n \rightarrow 3n \rightarrow 3n - n \rightarrow 2n + 8 \rightarrow \dfrac{2n + 8}{2} = n + 4$

d) If you have 2n and then add 8, the result is $2n + 8$.

11. ━ → ━ ━ ━ → ━ ━ → ━ ━ → ━ ━ ▪▪▪▪▪▪ → ━ ▪▪▪▪

13. They make a conjecture from examples (induction) and then they prove it (deduction).

15. a) Yes.

    b) If you are sick, then you have a fever. (c) No.

17. If a triangle has two equal sides, then it has two equal angles. If a triangle has two equal angles, then it has two equal sides.

19. $2x = 4$ if and only if $x = 2$.

21. Use enough steps to make it deceptive. An example would be the following. Subtract 3. Multiply by 2. Subtract 4.

    Divide by 2. Subtract your original number. Add 7.

23. a) An even number.

    b) The sum of the two even numbers is $2m + 2n$.

    $2m + 2n = 2(m + n)$

    $2(m + n)$ is an even number because 2 times any whole number is even.

25. a) If $x + 3 \neq 5$, then $x \neq 2$.

    b) If I don't confess, then I am not guilty.

    c) Possible: True statement: If I am 6 feet tall, then I am over 5 feet tall.

    Inverse: If I am not 6 feet tall, then I am not over 5 feet tall.

27. a) If $x \neq 2$, then $x + 3 \neq 5$.

    b) If I am not guilty, then I do not confess.

    c) Possible. If I like flowers, then I like roses.

    If I don't like roses, then I don't like flowers.

29. a) If I drove to work, then it was raining.

    b) If I did not drive to work, then it was not raining.

    c) If it was not raining, then I did not drive to work.

    d) Part (b).

31. c)

33. No.

35.

A	B	C	D
5	16	3	10
E	F	G	H
4	9	6	15
I	J	K	L
14	7	12	1
M	N	O	P
11	2	13	8

1.4 Lesson Exercises

LE1   Opener

Desks are placed in rows.

LE2   Concept

The sequences are arithmetic sequences.

LE3   Concept

a) Arithmetic. Difference = 2.

b) Not arithmetic.

c) Add 2.

LE4   Skill

a) 12, 10, 8, 6…

b) Yes.

c) FIRST = 12   NEXT = PREVIOUS - 2

d) $14 - 2(5) = 4$

LE5   Skill

a) 3

b) 5, 7, 9

LE6   Reasoning

   a) $a+3d, a+4d$

   b) $a+5d$

   c) $a+9d$

   d) $a+(n-1)d$

LE7   Reasoning

   a) 125, 135, 145, 155

   b) Yes.

   c) $125+(n-1)10$

   d) $125+(50-1)10 = 125+490$
   $$= 615$$

   e) Inductive

   f) Deductive

LE8   Concept

   No, the amount of increase must be the same for the sequence to be an arithmetic sequence.

LE9   Concept

   The geometric sequence as a common ration r between each pair of consecutive items.

LE10   Skill

   a) (i) neither   (ii) arithmetic   (iii) geometric

   b) (ii) a = 12   d = 8   (iii) a = 100   r = 1/5

   c) (ii) FIRST = 12   (iii) FIRST = 100

       NEXT = PREVIOUS − 8   NEXT = PREVIOUS x 1/5

LE11   Concept

   a) $a, ar, ar^2, ar^3, ar^4, ar^5$

   b) ar^5

   c) ar^9

   d) ar^{n-1}

LE12 Concept

a) 40, 120, 360

b) Geometric

c) FIRST = 40   NEXT = 3 x PREVIOUS

d) $40 \cdot 3^{n-1}$

e) $40 \cdot 3^{79}$

f) Inductive

g) Deductive

LE13 Concept

a) Sequence of squares

b) Neither

LE14 Reasoning

a) $1 = 1^2$ $1 + 3 = 2^2$ $1 + 3 + 5 = 3^2$

b) $1 = 1^2$ $1 + 3 = 2^2$ $1 + 3 + 5 = 3^2$

c) $1 + 3 + 5 + 7 = 4^2$, true

d) 3

e) 4

f) The sum of the first n odd numbers is n^2.

g) inductive

h) $40^2 = 1600$

LE15 Reasoning

a) $5^2 - 4^2 = 9$ Yes, it's true.

b) $c^2 - (c-1)^2 = c + (c-1)$

c) Does $c^2 - (c-1)^2 = c + (c-1)$?

$2c - 1 = 2c - 1$

d) inductive, deductive

LE16 Reasoning

All odd numbers except 1, $3 = 2^2 - 1^2$, and all multiples of 4 except 4, $8 = 3^2 - 1^2$.

LE17 Summary

Sequences are many times classified as arithmetic, geometric. An arithmetic sequence, 2, 7, 12, 17, ... has a common difference between terms, in this example the difference is 5. A geometric sequence, 4, 16, 64, 256…has a common ratio between terms, in this example the ratio is 4.

1.4 Homework Exercises

1.   a) not arithmetic

b) Arithmetic   difference $= -30$

c) Multiplication by 2.

3.   a) 60. 100, 140, 180,…

b) Yes, arithmetic

c) FIRST $= 60$ NEXT $=$ PREVIOUS $+ 40$

5. $n = 1,\ 10 - 1 = 9$

$n = 2,\ 10 - 2 = 8$

$n = 3,\ 10 - 3 = 7$

$n = 4,\ 10 - 4 = 6$

$n = 5,\ 10 - 5 = 5$

7.   a) Yes.

      b) FIRST = 2   NEXT = PREVIOUS + 7

      c) 2, 9, 16, 23,...

      $a + (n-1)d = 2 + (n-1)7 = 7n - 5$ 10th term $= 7(10) - 5 = 65$

      d) 100th term $= 7(100) - 5 = 695$

      e) nth term $= 7n - 5$

9.   a) $580 + (n-1) - 6 = 580 - 6n + 6 = 586 - 6n$

      b) $586 - 6(30) = 586 - 180 = 406$

      c) Inductive reasoning

      d) Deductive reasoning

      e) Does $100 = 586 - 6n$? Yes if $n = 81$, so 100 is a term in the sequence. deductive reasoning

11.   a) 0, 2, 4, 6, 8,...

      b) $2n - 2$

      c) FIRST = 0   NEXT = PREVIOUS + 2

13.   a) $1 + 49(2) = 99$

      b) $1 + 3 + 5 + ... + 99 = 100 \times \dfrac{50}{2} = 2500$

15.   a) 14

      b) $8 + 2(n-1) = 2n + 6$

17.   a) nth term $= a + (n-1)d$ where a is the initial term, n is the term number and d is the common difference

      b) a represents the initial term. Each successive term can be obtained by adding the common difference between terms.

19.   a) (i) arithmetic   (ii) neither   (iii) geometric

      b) (i) a = 50 d = 11   (iii) a = 600 d = 1/5

      c) (i) FIRST = 50   NEXT = PREVIOUS + 11   (iii) FIRST = 600   NEXT = PREVIOUS x 1/5

21.   a) This is approximately a geometric sequence with $r \approx 1.41$

      b) $a = 1, r \approx 1.41$

23. a)

Time (years)	0	5600	11,200	16,800	22,400
Fraction of C-14 left	1	1/2	1/4	1/8	1/16

b) About 13,000 years (since 1/5 is between 1/4 and 1/8).

25. a) The nth term for the arithmetic sequence is $3+12(n-1)$. The 30th term would be $3+12(30-1)=3+12*29$
$$=351$$

b) The nth term for the geometric sequence is $3 \bullet 12^{(n-1)}$. The 30th term would be $3 \bullet 12^{(30-1)} = 3 \bullet 12^{29}$

27. a) geometric

b) n^{th} term $= 5(7)^{n-1}$

c) $5(7)^{39}$

29. a) arithmetic

b) $30 + (n-1)(-10) = 40 - 10n$

c) -560

31. a) 27

b) 3, 9, 15, 21, 27,..., 159

Because its arithmetic, $3 + (n-1)6 = 6n - 3$

$$a + (n-1)d$$

Find out how many terms are in the sequence: $6n - 3 = 159$

$$6n = 162$$

$$n = 27 \text{ terms in the sequence}$$

33. a) 7 or 8

b) For 7: $1, ^{+1} 2, ^{+2} 4, ^{+3} 7,...$ (n^{th} term = previous term + n − 1)

For 8: n^{th} term $= 2^{n-1}$

35. a) 1, 1, 2, 3, 5, 8, 13, 21, 34, 55

b) The sum is 1 less than the term.

c) The sum of the first n terms is 1 less than the (n + 2) term.

d) The sum of the first 5 terms $= 1 + 1 + 2 + 3 + 5 = 12$, which is 1 less than the 13, the 7th term.

37.   a) There was another planet that exploded and became asteroids.

b) $4 + (3 \times 2^6) = 196$

c) Yes.

d) It is located where the missing planet should be.

e) $4 + (3 \times 2^7) = 388$; not so close.

39.   (b)

41.   a)

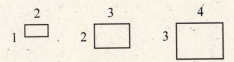

b) 3×4

c) 4×5

d) The sum of the first n even numbers is $n \times (n + 1)$.

e) Inductive

f) $31 \times 32 = 992$

43.   a) $5^2 - 3^2 = 16$ yes

b) $c^2 - (c - 2)^2 = 4(c - 1)$

c) Does $c^2 - (c - 2)^2 = 4(c - 1)$?

$c^2 - c^2 + 4c - 4 = 4c - 4$?

$4c - 4 = 4c - 4$

45.	a) $1 + 8 \times 10 = 9^2$ Yes.

b) $1 + 8\left(\dfrac{n(n+1)}{2}\right) = (2n+1)^2$

c) Does $1 + 8\left(\dfrac{n(n+1)}{2}\right) = (2n+1)^2$?

$1 + 8\left(\dfrac{n^2+n}{2}\right) = (n+1)(n+1)$?

$1 + 4n^2 + 4n = 4n^2 + 4n + 1$?

$4n^2 + 4n + 1 = 4n^2 + 4n + 1$

47.	a) 1

b) 3

c) 12

49.	No. Original sequence = 	2, 4, 8, 16,…

Original sequence + 1 = 3, 5, 9, 17,…

51.	The height of each bounce decreases geometrically.

53.	101 (sum) x $\dfrac{100}{2}$ (pairs) = 5050

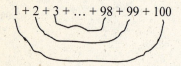

$1 + 2 + 3 + \ldots + 98 + 99 + 100$

55.	a) $y = 21$

b) $y = \dfrac{n(n+1)}{2}$

57. $15 + 16 + 17 + \ldots + a + (n - 1)d$

$a + (n - 1)d =$

$15 + (n - 1)1 =$

$15 + n - 1 =$

$n + 14$

So when n = 15 then 15 + 14 = 29 = last term. So first term + last term = 15 + 29 = 44,

So $15 + 16 + 17 + \ldots + 29 = 44\left(\dfrac{15}{2}\right) = \dfrac{660}{2} = 330$

1.5 Lesson Exercises

LE1   Opener

For student.

LE2   Opener

9

LE3   Opener

120 x .25 = $30;   50 − 30 = $20 more to make. $20 ÷ .25 = 80 more apples

LE4   Opener

10 cows left.

LE5   Concept

5 x 6 = 30 and 30 − 2 = 28 children in class. Multistep translation problem.

LE6   Concept

For the student. Puzzle problem.

LE7   Concept

8 ÷ 4 = 2. One-step translation problem.

LE8   Concept

a) The current ages of Jamaal and his daughter.

b) Yes.

c) Yapper's age.

d) Let x = Jamaal's age today and y = his daughter's age today.

Then x − 22 = Jamaal's age 22 years ago, and y − 22 = his daughter's age 22 years ago.

$$y - 22 = \frac{1}{4}(x - 22) \text{ and } y = \frac{1}{2}x$$

$$\frac{1}{2}x - 22 = \frac{1}{4}(x - 22)$$

$$\frac{1}{4}x = 66/4$$

$$x = 66, \quad y = 33$$

LE9   Reasoning

a) <u>Understanding the problem</u>: What are you supposed to find? (The total number of squares of all sizes in the picture.)

<u>Devising a plan</u>: How will you do this? (Group the squares by size. Count all the 1 by 1 squares. Then count all the 2 by 2 squares. Finally count all the 3 by 3 squares.)

<u>Carrying out the plan</u>: carry out your plan and obtain an answer.

Size	Number of squares
1 by 1	9
2 by 2	4
3 by 3	1
Total	14

<u>Looking back</u>: Make up another problem like this one. How many different squares are in the following picture?

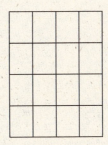

After solving this problem and the preceding one, develop a general solution for solving the same kind of problem with any size square.

b) $1^2 + 2^2 + 3^2 + 4^2 = 30$ squares

c) $1^2 + 2^2 + 3^2 + 4^2 + 5^2 = 55$ squares

d) Total number of squares in a square size n x n is $1^2 + 2^2 + 3^2 + \ldots + n^2$

e) $1^2 + 2^2 + 3^2 + 4^2 + 5^2 + 6^2 + 7^2 + 8^2 = 204$ squares

LE10 Reasoning

  a) $1 + 3 + 5 + 7 + 5 + 3 + 1 = 25$ dots

  b) For the student

LE11 Reasoning

  a)

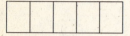

  (i) 5 top, 5 bottom, 6 vertical = 16 toothpicks

  (ii) 4 toothpicks in the first block, 3 toothpicks in each of 4 successive blocks, so $4 + 4 \times 3 = 4 + 12 = 16$

     toothpicks

  (iii) $(4 \times 5) - 4 = 16$ toothpicks

LE12 Concept

  Show two methods for finding the sum of five consecutive numbers whose sum is 40. $(6 + 7 + 8 + 9 + 10)$

LE13 Summary

  Polya's 4 step problem solving method:

  1. Understand the problem.

  2. Devise a plan.

  3. Carry out the plan.

  4. Look back.

1.5 Homework Exercises

1.   Multistep translation

3.   You place an order for 3 servings of pond water and 1 serving of twice-baked kelp. What is your total bill?

5.   Polya's four steps for problem solving are understand, plan, solve, and check. First read the problem and

    make sure you understand all of the information. Next devise a plan to solve the problem. Then carry out the

    plan. Finally, see if your results make sense.

7.

$x \cdot x = 49$ sq m

$x^2 = 49$ sq m

$x = 7$ m   length of fence $= 4 \cdot 7 = 28$ m

9.   Pull out 5 socks to be sure of getting a matching pair. Understand: How many socks must I pull out to get a matching pair? Plan: try pulling out socks and see the result. Solve: you need 5 socks.

11.   4 cuts. Understand: How are the cuts made? Plan: Make the cuts and see the results. Solve: 4 cuts.

13.   15   1 x 1 squares

       8   2 x 2 squares

       3   3 x 3 squares

       $15 + 8 + 3 = 26$

15.   a) 5 children

       b) $N + T$

17.

Move the two dots at the ends of the top row to the ends of the 3rd row. Then move the bottom dot to the top.

19.

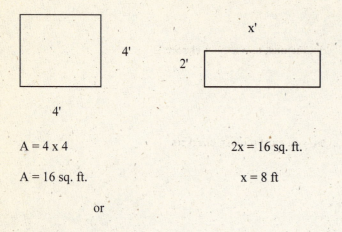

A = 4 x 4   2x = 16 sq. ft.

A = 16 sq. ft.   x = 8 ft

             or

       Need 16 boxes, so it must be 8' long.

21.   a) $4+5+6+7=22$

        b) $n+n+1+n+2+n+3=4n+6$

23.   Divide the coins into 3 groups: 6, 5 and 5. Put 5 coins on each side of the balance. If they are equal, clear the balance, go to the 6 coins and place 3 on each side of the balance. Choosing from the lighter side, clear the balance and put 1 on each side. If equal the remaining coin is the counterfeit. Otherwise the lighter coin is. If the 5 coins on each side of the balance are not equal, divide the coins from the lighter side into groups of 2, 2 and 1 and clear the balance. Place the two groups of 2 on the balance. If they are equal, the remaining is the counterfeit. Otherwise, choose the lighter side. Now place both coins on the balance, one on each side. The lighter side is the counterfeit

25.   a) 27

        b) 9

27.   a) Switch 2 with 7 and 4 with 9.

    b) 0

    c) 2

    d) For n pairs of checkers in which n is odd, you must move $n - 1$ checkers.

    e) no

    f) no

    g) For n pairs of checkers in which n is even, you must move n checkers.

    h) 24

    i) 50

1.6 Lesson Exercises

LE1   Opener

    a) Guessing and checking can lead to a correct solution.

    b) Making a table or list helps to organize data.

    c) Drawing a picture helps to clarify what needs to be done to solve a problem.

LE2   Reasoning

    a)

Oranges	Grapefruits	Total Cost	Actual Cost	Difference
10	8	$0.19(10) + 0.29(8) = \$4.22$	4.62	$0.40

    b)

Oranges	Grapefruits	Total Cost	Actual Cost	Difference
10	8	$0.19(10) + 0.29(8) = \$4.22$	4.62	$0.40
9	9	$0.19(9) + 0.29(9) = \$4.32$	4.62	0.30
8	10	$0.19(8) + 0.29(10) = \$4.42$	4.62	0.20
7	11	$0.19(7) + 0.29(11) = \$4.52$	4.62	0.10
6	12	$0.19(6) + 0.29(12) = \$4.62$	4.62	0

LE3   Reasoning

a)

Number of Gallons	Number of Quarts	Total Cost
0	11	$38.50
1	7	$33.00
2	3	$27.50
3	0	$25.50

b) Three gallons would be least expensive.

LE4   Reasoning

a) It would take around 30 days.

b)

$3 + (n - 1)1 = 30$ feet

$3 + n - 1 = 30$

$n + 2 = 30$

$n = 28$ days

LE5   Reasoning

$4 + (n - 1)3 = 30$

$4 + 3n - 3 = 30$

$3n + 1 = 30$

$3n = 29$

$n = 9\frac{2}{3}$ or 10 days

LE6   Reasoning

$$4 + (n - 1)2 = 50$$

$$4 + 2n - 2 = 50$$

$$2n + 2 = 50$$

$$2n = 48$$

$$n = 24 \text{ days}$$

LE 7   Reasoning

a) A well is 80 ft deep. A snail climbs up 7 ft each day but slides back 3 ft each night. On what day does he reach the top of the well?

b) A well is 80 ft deep. A snail climbs up 8 ft each day but slides back 4 ft each night. On what day does he reach the top of the well?

c) A well is 80 ft deep. A snail climbs up 12 ft each day but slides back 8 ft each night. On what day does he reach the top of the well?

LE8   Reasoning

Number of Players (each team)	Number of Handshakes
1	1
2	6
3	10
4	28
5	45
6	66
7	91

LE9   Skill

a) If you estimate the fish at $4, the OJ at $3, and the mushrooms at $2, a total of $9, less than the $10 Veronica has, so she has enough money.

b) How much change would Veronica receive from a $10 bill?

LE10 a) The temperature and stopping for lunch is not useful information.

b) You need to determine how many hours of driving occurred.

LE11 Summary

Problem solving strategies offer a variety of ways to represent the information.

1.6 Homework Exercises

1. In the guess and check strategy, you start by making a guess. Then check to see how accurate it is. Make a

new guess using what you learned from your check. Continue guessing and checking until you find the

answer.

3. $x(x + 5) = 1184$ $x = $ Linda $x + 5 = $ Craig

$x^2 + 5x - 1184 = 0$ $x = 32$ $x + 5 = 37$

$(x + 37)(x - 32) = 0$

$x + 37 = 0$ $x - 32 = 0$

$x = -37$ $x = 32$

5.

Oranges	Apples	Total Cost	Actual Cost	Difference
14	11	$0.39(14) + 0.24(11) = \$8.10$	8.40	$0.30
15	10	$0.39(15) + 0.24(10) = \$8.25$	8.40	0.15
16	9	$0.39(16) + 0.24(9) = \$8.40$	8.40	0

7.

	B	D	F	GE	HD	R
B		4	4	4	4	4
D			4	4	4	4
F				4	4	4
GE					4	4
HD						4
R						

$4 + 8 + 12 + 16 + 20 = 60$ games

9.

	Manager	Assistant Manager	Secretary 1	Secretary 2	Sales Agent 1	Sales Agent 2
Week 1		X			X	
Week 2		X			X	
Week 3			X			X
Week 4			X			X
Week 5	X			X		
Week 6	X			X		
Week 7	X					

11. Day   Height

1   5
2   5 + 3
3   5 + 2(3)
.
.
.
n $5 + (n - 1)3 = 3n + 2$
$3n + 2 = 100$
$3n = 98$
$n = 32⅔$ or 33 days

13. a) possible answer: A worm goes up 4 feet and slides back 1 foot each day.

b) possible answer: A worm goes up 6 feet and slides back 3 feet each day.

c) possible answer: A worm goes up 9 feet and slides back 3 feet each day.

15. a) 2 people → 1 handshake

b) 3 people → 1 + 2 = 3 handshakes

c) 4 people → 1 + 2 + 3 = 6 handshakes

d) 8 people → 1 + 2 + ... + 7 = 28 handshakes

17. For the student.

19. Guess and check.

21. Draw a picture

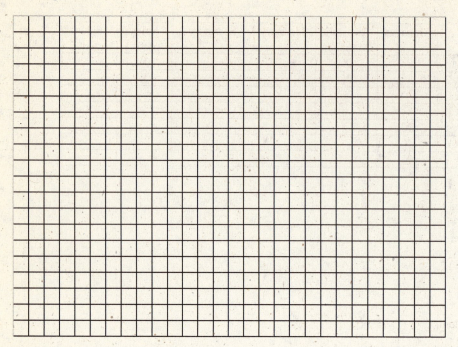

Each grid is 6×6, the dimension of the tiles. The region is 14 feet by 10 feet. There are 28 tiles horizontally and 20 tiles vertically, so a total of 560 tiles.

23. 17 is the impossible score. Make a table.

25. If we round 76 to 80 and 24 to 25, then the product of 80 and 25 is 2000. Since each value was rounded up, the weight limit is not exceeded with this approximation.

27. a) The phone service cost for last month is not needed.

b) We need to know how many calls Claudinna makes this month.

29. a) 40 yds. (a square)

b) $4\sqrt{N}$ yds.

31. First, he takes the goose across. Then he takes the fox or corn across and picks up the goose when he drops off the fox or corn. He brings the goose back to the starting side and picks up the corn or fox and takes that to the other side. Finally he returns and brings the goose across again.

33 a) 9 tables

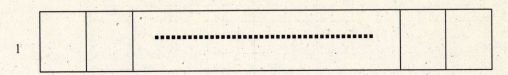

b) $n - 1$ tables.

35. Yes, they are equivalent.

$$6 + 4(T - 1) \overset{?}{=} 2 + 4T$$
$$6 + 4T - 4 \overset{?}{=} 2 + 4T$$
$$2 + 4T = 2 + 4T$$

37. Left for the student.

Chapter 1 Review Exercises

1. Inductive reasoning is a form of reasoning in which many specific cases are observed, and a generalization is made based on the patterns observed in the specific cases. Many times it leads to a true conclusion; however, sometimes it does not.

3. Yes. $1 + 2 + 3 = 6$ and 3 divides 6. $7 + 8 + 9 = 24$ and 3 divides 24.

5. The process of reaching a necessary conclusion from given hypotheses.

7. a) You did not finish your fish.

b) Unknown

9. Induction

11. Deduction

13. $n \rightarrow n + 5 \rightarrow 3n + 15 \rightarrow 3n + 6 \rightarrow 2n + 6 \rightarrow n + 3$

15. a) If it has four congruent sides, then the rectangle is a square.

b) If I call, then there is a problem.

17. a) arithmetic

b) $a + (n - 1)d = 20 + (n - 1)(-1)$

   $= 20 - n + 1$

   $= 21 - n$

c) $21 - 40 = -19$

d) Inductive

e) Deductive

19. a) neither

b) N/A

c) N/A

d) 3, 12, 27, 48, 75, 108   If n > 1, the nth term = previous term + 3(2n − 1). Another way to look at it is you start with 3, add 9 to get 12, add 9 + 6 = 15 to 12 to get 27, add 15 + 6 = 21 to 27 to get 48, add 21 + 6 = 27 to 48 to get 75, add 27 + 6 = 33 to 75 to get 108, and so on.

21. a) $7^2 - 4^2 = 3 \bullet 11$

b) $n^2 - (n - 3)^2 = 3(2n - 3)$

c) $n^2 - (n - 3)^2 \overset{?}{=} 3(2n - 3)$
   $n^2 - (n2 - 6n + 9) \overset{?}{=} 6n - 9$
   $6n - 9 = 6n - 9$

23. First, the mechanic would find out what was wrong with the car. Next, the mechanic devise a plan to fix the problem. Then the mechanic would try out the repair plan. Finally the mechanic would check to see if the problem no longer occurred.

25.   x   x   x

width	length	number of rectangles
1	x	6
1	2x	4
1	3x	2
2	x	3
2	2x	2
2	3x	1

$6 + 4 + 2 + 3 + 2 + 1 = 18$ rectangles

27. Split the coins into $3 - 3 - 3$. If the $3 - 3$ are even on the balance, the remaining 3 are split $1 - 1 - 1$. If the scale is balanced, the remaining one is the counterfeit. If the scale is not balanced, the lighter side is the counterfeit. If the $3 - 3$ are not even on the balance, choose the lighter side and split $1 - 1 - 1$. Now you can determine the counterfeit. Put $1 - 1$ on the balance. If even, the remaining coin is the counterfeit, otherwise the lighter side is the counterfeit.

29. $\dfrac{4 \bullet 3}{2} = 6$ 4 choices for the first person x 3 choices for the second person divided by the number of ways to arrange 2 people (2! or 2)

31. a) (possible answer) A well is 40 ft deep. A snail goes up 7 ft and slides back 5 ft each day. On what day does it reach the top?

b) (possible answer) A well is 40 ft deep. A snail goes up 5 ft and slides back 3 ft each day. On what day does it reach the top?

Chapter 2

2.1 Lesson Exercises

LE1   Opener

    A set is a collection of objects.

LE2   Concept

    a) $\in$

    b) $\notin$

LE3   Concept

    The set of positive numbers less than 0.

LE4   Concept

    a) A, B, D, and F are equivalent, C and E are equivalent.

    b) $A = F$

    c) No

    d) yes

    e)

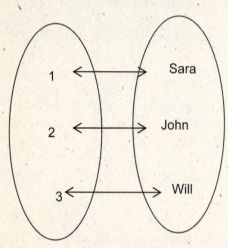

    f) Put E in a one – to – one correspondence with set C.

LE5   Concept

    a) Finite

    b) Finite

    c) Infinite

LE6   Concept

a) {Sunday, Monday, Wednesday, Friday}

b) The days you do not exercise.

LE7   Concept

a, d

LE8   Concept

a) $\in$

b) $\subseteq$

LE9   Reasoning

a) Observe a pattern between the number of elements in a set and the number of subsets.

b)

Set	List of Subsets	Number of Subsets
{1}	{} {1}	2
{1, 2}	{1} {2} {1, 2} Φ	4
{1, 2, 3}	{1} {2} {3} {1, 2} {1, 3} {2, 3} {1, 2, 3} Φ	8
{1, 2, 3, 4}	{1} {2} {3} {4} {1, 2} {1, 3} {1, 4} {2, 3} {2, 4} {3, 4} {1, 2, 3} {1, 2, 4} {2, 3, 4} {1, 3, 4} (1, 2, 3, 4} Φ	16

c) Now 4 can be matched up with each subset from {1, 2, 3}, thus doubling the number of subsets.

d) 2^n

e) Inductive

LE10   Summary

$\in$ stands for element or member of a set

$\subseteq$ stands for subset of a set

Every set has 2^n subsets

2.1 Homework Exercises

1.   a) T

         b) F

         c) T

         d) F

3.   a and b

5.   D and F, E and N

7.   a) False

         b) $A = \{1, 2, 3\}$ $B = \{a, b, c\}$ $A \neq B$ but A is equivalent to B

9.   a) Infinite

         b) Finite

11. $\overline{A} = \{\text{Spanish, Science}\}$ $\overline{B} = \{\text{History}\}$

13.

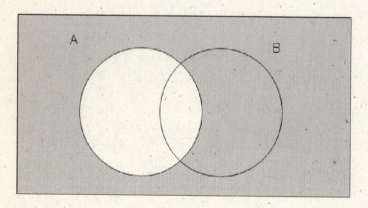

15.   a, c, d

17.   a) T

         b) F

         c) T

         d) T

19.   a) $\subseteq$

         b) $\in$

21.   {B, R, J}{B, R, M}{B, S, J}{B, S, M}{B, J, M}{R, S, J}{R, S, M}{R, J, M}{S, J, M}

        {B, R, S, M}{B, R, J, M}{B, S, J, M}{R, S, J, M}{B, R, S, J, M}

23.   a) Match up each number with twice that number

        b) Match up each number with 10 more that that number

25.   a) 3

        b) 5

        c) 7

        d) $2n-1$

27. $kn-k+1$

2.2 Lesson Exercises

LE1   Opener

        They are members of both math club and science club

LE2   Concept

        a) $A \cap B = \{Li, Clara\}$

        b)

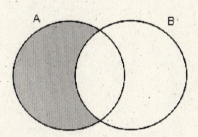

        Students in math club but not in science club

        c) $A \cap \overline{B}$

LE3   Reasoning

        $C \subseteq B$

LE4   Concept

        a) {Joe, Sam, Li, Clara, Juan}

        b) At a joint meeting of the math and science clubs

LE5   Reasoning

a) True

b) False

c) True

LE6   Skill

a) {20, 30}

b) {1, 10, 20, 30, 100}

c) {1, 10, 20, 30, 100}

$$A \cup (B \cap C) = (A \cup B) \cap (A \cup C)$$

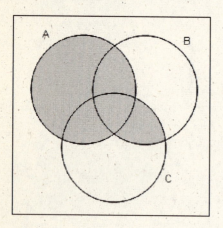

$A \cup (B \cap C)$

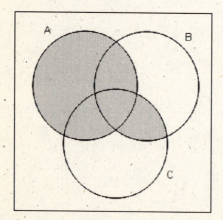

$(A \cup B) \cap (A \cup C)$

LE7   Reasoning

a) and b)

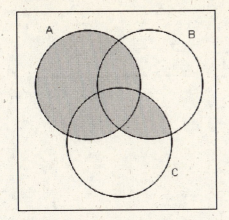

c) $(A \cup B) \cap (A \cup C) = A \cup (B \cap C)$

$(1, 2, 3, 4, 5, 6) \cap (1, 2, 4, 5, 6) = (1, 2, 4, 5, 6)$

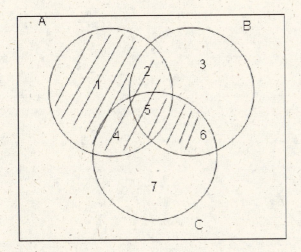

LE8   Concept

a) $\cup$

b) $\in$

c) $\subseteq$

LE9   Concept

This is not correct because there are college graduates that are not in the teacher section.

LE10   Concept

a)

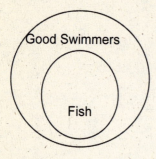

b)

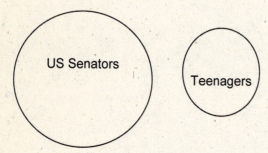

c)

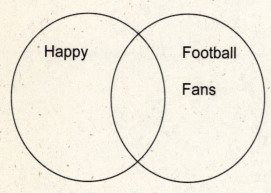

LE11   Concept

a) No animals are plants

b) $A \cap P = \phi$

LE12   Reasoning

a) 800

b) 300

c) 650

LE13   Reasoning

   a) Candice, Dan, Gwen

   b) Brian, Brad, Celeste, Mike, Sara

   c) Brian

LE14   Reasoning

   a)

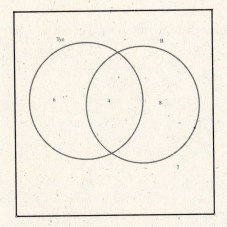

   b) 7

   c) and d)

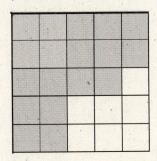

LE15   Reasoning

   a) 12

   b) 0

LE16   Concept

   Small and large blue squares

LE17   Reasoning

   The large gray, blue and white triangles

LE18   Reasoning

       1 = {big blue and white triangles}

       2 = {small gray triangles}

       3 = {big gray triangles}

       4 = {big gray square, circles and rectangles}

LE19   Summary

Intersection and union are convenient methods for describing characteristics shared or not by objects under question.

2.2 Homework Exercises

1.   a) {3, 9}

      b) {1, 3, 5, 6, 7, 9, 11, 12, 15, 18}

      c) no

3.   The two streets pass through the intersection, which is the region that is common to both streets

5.   a) true

      b) False

      c) False

7.   a) The set of female education majors

      b) The set of college students 22 years or older who are not education majors

      c) Set U

      d) { }

9.   Fish or spinach or both

11. $B \subseteq C$

13. a) $(A \cup B) \cup C = A \cup (B \cup C)$ is associative

 b)

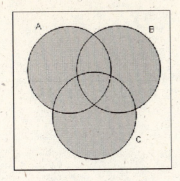

$(A \cup B) \cup C$

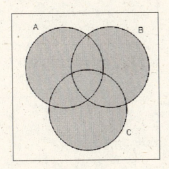

$A \cup (B \cup C)$

15.   a) {2, 3}

        b) {2, 3, 4, 6, 8}

        c) {2, 3}

        d) a and c

        e)

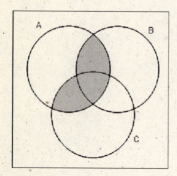

$A \cap (B \cup C)$

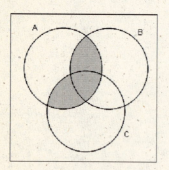

$(A \cap B) \cup (A \cap C)$

17.   a) 5

      b) 3

19.   a)

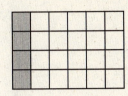

      b)

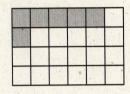

      c) 16

21.   a) 22

      b) $25 + 22 = 47 - 40 = 7$

23.   a) T

      b) T

      c) T

      d) F

25.   a) $\in$

      b) $\cap$

      c) $\subseteq$

27.   a) Impossible

      b) $A = \{1\}$ $B = \{2\}$

      c) $A = \{1\}$ $B = \{1\}$

29.   a)

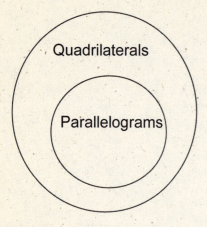

$P \subseteq Q$

b)

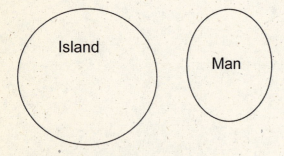

$M \cap I = \{\}$

c)

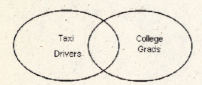

31.   a) All tomatoes are fruits

        b) $T \subseteq F$

33.   (e)

35.   a) 125

        b) 350

        c) 10

        d) a) $T \cap R = 125$
           b) $T \cap \overline{R}$
           c) $\overline{T} \cap \overline{R}$ or $\overline{T \cup R}$

37.   a)

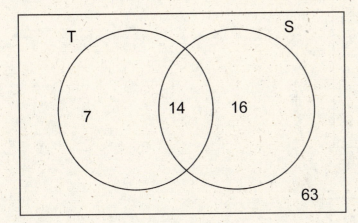

        b) 63

39.   For student

41.   a) For student

        b) 4, 5

        c) 2, 3

        d) 1, 2, 4, 5

        e) 6

43.   a) This is the set of all triangles and gray attribute blocks.

        b) This is the set of all gray triangles.

45.   a) All gray blocks and triangle blocks

        b) Gray triangles

47.   I. small rectangles that are not blue

II. small blue shapes that are not rectangles

III. small blue rectangles

IV. blue rectangles that are not small

49.   (c) polygons with the interior shaded.

51.   a) $58 + 14 + 56 + 8 = 136\%$

b) Some people put down more than one response

53.   a)

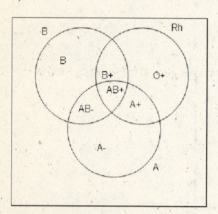

b)

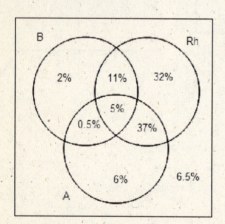

55.   a) $3 \times 2 = 6$

b) (yellow, red) (yellow, blue) (green, red) (green, blue)

57. $10 \times 7 = 70$

59.   (b) $U = \{1, 2, 3, 4, 5, 6, 7, 8\}$

$A = \{1, 2, 3\}$

$B = \{3, 4, 5\}$

$\overline{A \cup B} = \{6, 7, 8\}$

$\overline{A} \cap \overline{B} = \{4, 5, 6, 7, 8\} \cap \{1, 2, 6, 7, 8\}$

$= \{6, 7, 8\}$

61.   For student

2.3 Lesson Exercises

LE1   Opener

a) Temperature

b) Your clothing changes according to the temperature

c) The temperature is changing, it is dropping, the quantity is changing, it is becoming smaller

LE2   Concept

a)

Number of hours x	0	2	4	6	8
Pay in $ y	0	28	56	84	112

b) $y = 14x$

c) To find the dollars on pay (y) multiply the number of hours (x) by 14

d)

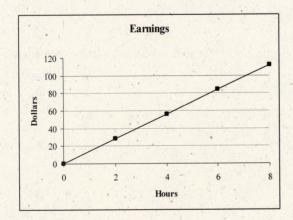

LE3   Concept

Domain (Input Set) = $\{0, 2, 4, 6, 8\}$

Range (Output Set) = $\{0, 28, 56, 84, 112\}$

LE4   Concept

Only diagram (a).

LE5   Concept

a) Not a function, more than 1 residence

b) Function

c) Not a function, February can have 28 or 29 days

LE6   Concept

a) In a vertical line

b) If any vertical line intersects a graph at more than one point, the graph does not represent a function

LE7   Skill

a) Function

b) Function

c) Not a function

LE8   Concept

a) Arithmetic

b) $30 + (n-1) \bullet 10 = 30 + 10n - 10$
$$= 10n + 20$$

c) $y = 10x + 20$

d) There is $20 plus $10 for each hour or the charge is $30 for the first hour and $10 for each additional hour.

$y equals $20 plus $10 for each hour (n)

LE9   Reasoning

a) 4

b) 0

c) For each increment of 1 hour , length decreases 2 inches

d) $H = 10 - 2T$

e)

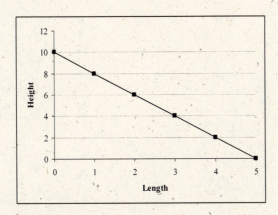

f) As T increases H decreases

LE10   Reasoning

a) $y = x^2 + 3$

b) If x is even, $y = 0$

If x is odd, $y = 1$

LE11   Summary

A function can be represented with a table, an equation, a rule in words, or a graph.

2.3 Homework Exercises

1.   a)

Number of hours x	0	2	4	6	8
Pay y	0	46	92	138	184

b) $y = 23x$

c) To find y, multiply x by 23

d)

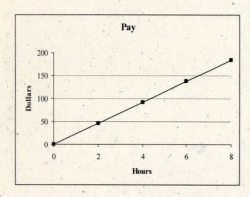

3.   Domain = {0, 2, 4, 6, 8}   Range = {0, 46, 92, 138, 184}

5.   a) The set of feet measured

      b) The measurements between 4 and 35 cm

7.   a) 1 and 3

      b) (1) squaring (2) "is a friend of " (3) "wants to be a"

9.   b and c

11.   b and c

13.   a) no

      b) yes

      c) no

15.   a) $500 \le F \le 800$

      b) $92,500 \le C \le 112,000$

      c) Each F value in the domain has exactly one C value in the range

17.   a) $P = 2n - 1$

        b) 9

        c) Inductive

        d)

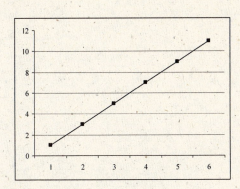

        e) The points all lie on a line

        f) $a + (n-1)d = 1 + (n-2)2 = 2n - 1$

19.   a) $V = L^3$

        b) 125

        c) n^{th} term $= n^3$

21.   a) $N = 12\sqrt{S}$

        b) 96 $n = 12\sqrt{64} = 12 \bullet 8 = 96$

23.   a) $y = 8$

        b) $y = 10$

        c)

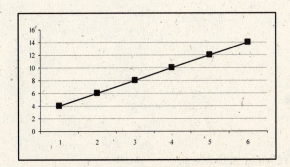

        d) The points lie on a line

e) $y = 2n + 2$

25. $3 \rightarrow 5$

         $N \rightarrow 8 - N$

27.   "Is taller than"   "Is heavier than"

29.   {(5, 9), (5, 18), (10, 18), (15, 18)}

31.   For the student

Review Exercise Chapter Two

1.   a) $2^3 = 8$

         b) A = {3, 4, 6} a set with 3 elements

         c)

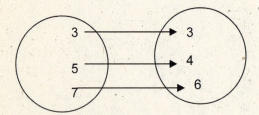

         d) {5, 7}

3.   a) 4

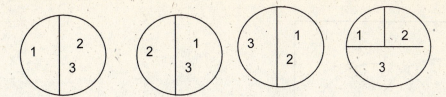

    b) 14

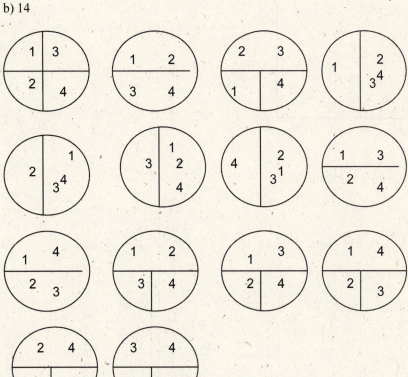

5.   a) T

       b) T

       c) T

7.   Intersection tells us everything that is in both sets. Union tells us what we would have if we put all the

items from both sets together.

9.   a) 5

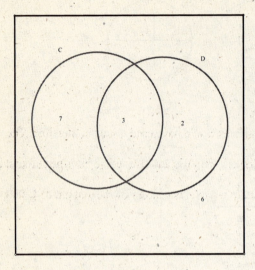

      b) Deductive

11.   Gray blocks that are not thick and not triangles

13.   Yes

15.   a) by 4

      b) 900

      c) $F = \dfrac{S^2}{4}$

      d) Weight, direction

      e) $0 < S \le 100 \text{ mph}$

17.   (d)

Chapter 3

3.1 Lesson Exercises

LE1   Opener

7 = VII in Roman numerals

LE2   Skill

a) These were symbols for familiar objects. The Egyptians would have used a staff in traveling. The scroll relates to their development of papyrus. The lotus flowers, tadpoles and fish would have been found in a culture that developed around the Nile river. The astonished man is amazed by the amount of $1,000,000$.

b)

LE3   Skill

LE4   Concept

It takes fewer distinct symbols and a smaller total number of symbols

LE5   Skill

$10 \cdot 60^2 + 31 = 36,031$

LE6   Skill

a) 13

b) $(20 \times 6) + 11 = 131$

c) $5 \cdot 20 = 100$

LE7   Skill

a) 1959

b) DCLXXII

LE8   Skill

d or f

LE9   Skill

$$100A + 10B + C$$

LE10   Communication

In a place value system, the value of a numeral changes depending upon its location. In a system that does

not have place value, the value of a numeral is always the same.

LE11   Skill

1030

LE12   Communication

When the child has ten units, it can be traded for a long (groups of 10). The child would now have 4 longs

or 4 groups of ten which is 40.

LE13   Skill

a) Count – number of letters counting unit letters

b) Count – number of seats, counting unit – seats

LE14   Skill

Measure – your height

Unit of measure – inches

LE15   Concept

a) count

b) measure

c) measure

LE16   Opener

a) 0, 1, 2, 3, or 4

b) 5, 6, 7, 8, or 9

LE17   Skill

a) 386,000

b)

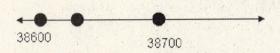

38600   38700

LE18   Reasoning

82,500 and 83,499 (inclusive) if attendance is rounded to the nearest thousand

LE19   Concept

a) 700

b) 627 is round up to the next 100

LE20   Summary

A place – value system reduces the number of different symbols needed

3.1 Homework Exercises

1.   a) 250

b) 1013

3.   a) Yes

b) Changing the position of different symbols in a Hindu Arabic numeral changes the value of the numeral.

Egyptian symbols can be written in any order.

5.   a)

        b)

        c)

7.   a) 9

        b) 120

        c) 247

9.   A symbol for zero, place value, 10 digits to represent all numbers

11.   a) 14

        b) 40

        c) 1613

        d) 1964

13.   a) 642

        b) 12

        c) 6

15.   a) $(4\times10^2)+(0\times10^1)+(7\times10^0)$ or $(4\times100)+(0\times10)+(7\times1)$

        b) $(3\times10^3)+(1\times10^2)+(2\times10^1)+(5\times10^0)$

17. $(A\times10^3)+(B\times10^2)+(C\times10^1)+(D\times10^0)$

19.   a) 1324

        b) 207

21.   a) 3 green 7 blue 2 yellow

        b) 1 green chip

        c) Chip trading is more abstract because chips for 1, 10, 100 and 1000 are all the same size. The value of a base – ten block is proportional to its size.

23.   1 hundred 2 tens 7 ones

        1 hundred 1 ten 17 ones

        1 hundred 0 tens 27 ones

25.   a) 3007

        b) Does not understand place value

        c) Go over the value of each place in place value

27.   a) measure

        b) count

        c) measure

29.   It's in between. Dollars are like a measure, but cents are like a count.

31.   a) 600

        b) $36,400

33.   a) Tens

        b) Hundreds

        c) Ten thousands

35.   a) $120

        b) Rounding down

37.   7500 and 8499 (inclusive)

39.   6,210,000,000

41.   For student

3.2 Lesson Exercises

LE1   Opener

        Addition is the number of elements in the union of two groups

LE2   Concept

      a) Yes

      b) induction

      c) no

LE3   Opener

Sue has 3 cookies and Sally has 5. How many cookies do they have together? Bob has been absent 3 days

one month and 5 days the next month. How many days of school have been missed altogether?

LE4   Concept

Combine groups

LE5   Concept

Combine measures

LE6   Connection

      a)

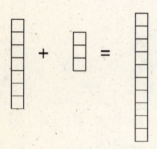

      b)

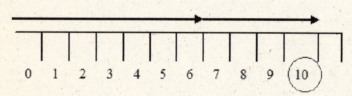

LE7   Skill

The child could count on, the child could use doubles plus three.

LE8   Concept

Blocks could be used. Start with 7 blocks and then count on 8, 9, 10

LE9   Skill

$2 + 4 = 6$

LE10   Opener

Dan has 6 pet fish. Two of them died. How many are left? Sally has 6 cups of flour. She gave two cups to Susan. How many cups are left?

LE11   Concept

$7 + 4 = 3$ Jaclyn has worked 3 hours longer. Compare measures

LE12   Concept

Take away a group

1 left

LE13   Skill

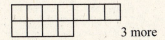 3 more

LE14   Communication

What do I need to add to 5 to get 13?

LE15   Reasoning

a)

b)

c) The sum of the far right and far left will always be 33

d) Inductive

e)

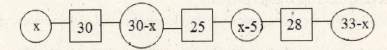

LE16   Reasoning

a)

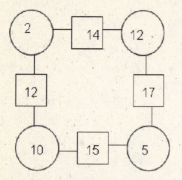

b)

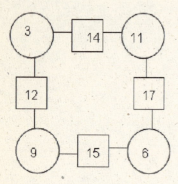

c)

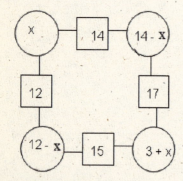

d) The sum of the opposite corners is always $2x + 3$

LE17   Summary

Addition applications that are beneficial for young students are combine groups and combine measures.

Beneficial subtraction applications are take – away groups and take – away measures, compare groups and compare measures.

3.2 Homework Exercises

1. (a)

3. 18 years

5. a) $2 + 3 = 5$

   b)

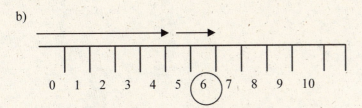

7. Joe has 4 marbles, Sam gives him 3, how many does he have now?

9. a) $2 + ? = 7$

   b) $37 + 72 = N$

   c) $n + 4 = x$

11. There are two fewer lids

13. 3 of the children play games

15. □ □ □ □ ⟶ □□□□□□ $7 - 4 = 3$

17. Take Away – Remove part of a group

    Compare – Two groups

    Missing Part – Know one part, figure out the other

19. a) $7 - 5 = 2$

    b)

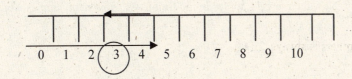

21. a) $8 - 5 = 3$

    b) Take away a group

23. a)

b)

c)

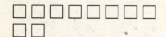

25. For the student

27. Julie has 5 cups of flour, she uses 2 cups for a recipe. How many cups does she have left?

29. a) When counting back one that would be 7 then 6, and 5

b) Give the child 8 blocks. Then have the child count off as they take away 3 from the set.

31. a) $10 - x$

b) $x - 2$

c) $x + 6$

33. $\$A - \$F - \$C + \$P$

35. a) $(a - b) \geq c$

b) Given whole numbers x, y, and z. Under what conditions would $(z - y) - x$ be a whole number?

37. a) Subtraction, missing part (groups)

b) Mary picked 5 apples today. Now there are 11 apples in the fruit bowl. How many apples were in the bowl before?

39. a) $8 > 6$ because $8 = 6 + 2$

b) For whole numbers a and b, $a < b$ whenever $a = b - k$ for some counting number k

41. a) $3 = 1 + 2$, $5 = 2 + 3$, $6 = 1 + 2 + 3$, $7 = 3 + 4$, $9 = 4 + 5$, $10 = 1 + 2 + 3 + 4$

b) Any counting number that is not a power of 2 can be written as the sum of two or more consecutive counting numbers.

c) Inductive

43.   The 2 should be subtracted from the 27, not added. The women first paid $30 and then got $3 back. So they

ended up paying $27, $2 to the bell person and $25 for the room.

45.   a)

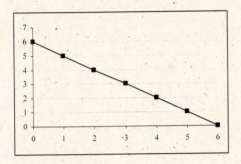

b) A straight line

c)

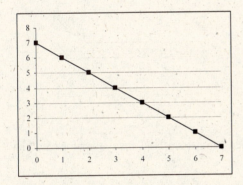

d) A straight line

e) If the ordered pairs of any fact family are plotted in a graph, the points will lie on a straight line.

f) Induction

47.   The two addends are the input and there is only one output the sum.

49.

4	3	8
9	5	1
2	7	6

3.3 Lesson Exercises

LE1   Opener

Multiplication is repeated addition

LE2   Skill

$3 \times 4 = 12$

LE3   Concept

a) Yes

b) Induction

c) No

LE4   Opener

I bought 3 packages of tomatoes. Each package contained 4 tomatoes. How many tomatoes did I have?

LE5   Connection

Equal groups (3 is repeated 4 times)

LE6   Connection

Area

LE7   Skill

$4 \times 3 = 12$

LE8   Skill

   $5 \times 4 = 20$

LE9   Skill

   $3 \times 4 = 12$

LE10   Skill

   $13 = 4 \times 3 + 1$

LE11   Opener

   I have a dozen eggs. How many 3 – egg omelets can be made? A walk is 12 miles long. How long will it

   take if a walker averages 3 miles per hour?

LE12   Connection

   Partition (share) a group

LE13   Connection

   Share measures

LE14   Connection

   Equal measures (How many 80's make 400?)

LE15   Skill

   $6 \div 3 = 2$

LE16   Skill

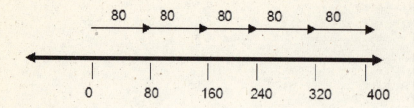

$400 \div 80 = 5$

LE17   Concept

$15 - 3 = 12, 12 - 3 = 9, 9 - 3 = 6, 6 - 3 = 3, 3 - 3 = 0$ Answer = 5, You can take away three 5 times so

$15 \div 3 = 5$

Repeated subtraction

LE18   Opener

$0 \overline{)4}^{\,c}$ No value of c will satisfy this problem

$0 \bullet c = 4$

LE19   Communication

a) $0 \div 4 = 0$

b) $0 \div 4 = ?$ is the same as $4 \times ? = 0$ so $? = 0$

LE20   Communication

a) $0 \div 0$ is undefined

b) $0 \div 0 = ?$ is the same as $0 \div ? = 0$, so ? could stand for any number. Since each division problem must

have one definite answer, we say $0 \div 0$ is undefined.

LE21   Summary

Multiplication is repeated addition, division is repeated subtraction, division by zero is not defined

LE22   Reasoning

a) $406 (double $200 and add $6)

b) Deductive

LE23   Reasoning

a) In class work.

b) $(7+3)\times 6 = 60$

c) $7+(3\times 6)=25$

d) 25 is correct

LE24   Skill

a) $6-6+7\times 3=21$

b) $12-9+10=13$

c)
$$m^2 - 4(n+2)$$
$$5^2 - 4(3+2) = 5^2 - 4(5)$$
$$= 25 - 4(5)$$
$$= 25 - 20$$
$$= 5$$

LE25   Skill

$$((3+3)-3)\div 3 = 1$$
$$(3\div 3)+(3\div 3) = 2$$
$$((3+3)+3)\div 3 = 3$$
$$3^{(3-3)}+3 = 4$$
$$(3+3)-(3\div 3) = 5$$
$$(3+3)+3-3 = 6$$
$$3+3+(3\div 3) = 7$$
$$(3\bullet 3)-(3\div 3) = 8$$
$$(3\bullet 3)+(3-3) = 9$$
$$(3\bullet 3)+(3\div 3) = 10$$

3.3 Homework Exercises

1.   a) 3×7

       b) 7, 14, 21

       c)

3.   a) Not closed $2 \times 2 = 4$, $4 \notin \{1, 2\}$

       b) Closed

       c) Closed

5.   Equal measures

7.   Equal groups

9.   Divide the interior in 1 ft squares as shown. It takes 6 squares to cover the interior, so its area is 6 sq. ft.

11. $3 \times 2 = 6$

13.   a)

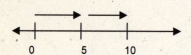

       b)

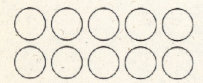

15.   A class has 8 rows of desks with 5 desks in each row. How many desks are there?

17.   A school is charging \$3 for lunch and \$6 for a concert. How much will it cost to buy 7 tickets for each event?

19. a) She is using the counting principle

   b) Tell her to take 2 counters, five times, the first factor is 5, tells how many times to take the 2nd factor of 2.

21. a) $8 \rightarrow 74$
   $x \rightarrow x^2 + 10$

   b) $10 \rightarrow 38$
   $x \rightarrow 4x - 2$

23. a) 3^2

   b) 4^3

25. a) $7 \times ? = 63$

   b) $N \times C = 0$

27. Partition equal measures

29. Equal measures

31. Equal groups

33. a) $8 \div 4 = 2$

   b)

35. a)

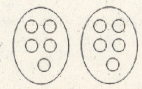

   b)

1				
2	2	2	2	2

   c)

37. Ten pieces of candy are to be divided among 5 children. How many pieces does each child receive?

39.   A package contains 10 barrettes. How many 2 barrette packages can I make?

41.   a) 9, 6, 3, 0

        b)

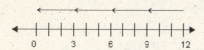

43.   a) Pens come in packs of 4. How many full packs can be made from 19 pens?

        b) Each car can hold 4 people, how many cars are needed for 19 people?

        c) In part a), how many pens are left?

45.   a) If he puts a maximum of 4 photos on a page, how many pages will he need?

        b) If he puts 4 photos on each page how many extra photos will be left over?

        c) If 4 photos fill a page, how many pages can he fill?

47.   a) 1

        b) 2

        c) Quotient

        d) Remainder

49.   a) 2 substitutes

        b) Remainder

51.   a) Undefined

        b) The expression $7 \div 0 = ?$ is the same as $? \times 0 = 7$ which has no answer. So we make $7 \div 0$ undefined.

53.   Determine if the child means 3 divided into 6 equal parts

55.   No. 3 is the divisor in the first example and the dividend in the 2nd example. The teacher should determine

        if the student meant 3 divided into 6 equal parts.

57.   a) How many 8's make 40? $8+8+8+8+8$

        b) How many 8's can you subtract from 40?

        $40 - 8 = 32, 32 - 8 = 24, 24 - 8 = 16, 16 - 8 = 8, 8 - 8 = 0$ 5 eights can be subtracted

        c) $? \times 8 = 40$ Since $? = 5$, $40 \div 8 = 5$

59. a) $\boxed{5}$

$\downarrow \times 4$

$\boxed{20}+6$

$\downarrow$

$\boxed{26}$

b) Deductive

61. Let x = # of daisies Frank picked

$$x - \frac{1}{2}x + 18 = x$$

$$\frac{1}{2}x + 18 = x$$

$$18 = \frac{1}{2}x$$

$$36 = x$$

63. Take out 1 chip. On successive turns, reduce the pot to 7 chips, 4 chips, and finally 1 chip.

65. a)

x	2	4
$x+3$	5	7

x	2	3
$4x$	8	12

	x	6	12
$4x+3$		27	51

b) Solved the equation for x: for ex. $x + 3 = 5$

$$x = 2$$

67. a) $8 - 4 + 6 = 10$

b) $49 + 18 = 67$

c) $x - (7 - y) + 3y$

$8 - (7 - 2) = 8 - 5$

$= 3$

69. a) 19, the result should be 5

b) 162

71. a) $r = 100$ b) $t = 3$ c) $y = 7$

73.   Whole number division is not closed since the quotient is not always going to be a whole number, for

              example 3 divided by 6 is one half, not a whole number.

75.   a) Multiplication, array and subtraction, take away groups

              b) Multiplication, equal groups and addition, combine groups

77.   a) Addition, combine groups and subtraction, take away a group

              $(10-(2+3))$ or both steps are subtraction, take away a group $(10-2$ and $8-3)$

              b) Multiplication, array and subtraction, take away a group

79.   For each set of (input) factors there is only one product (output).

81. $4n-4$

83.   Pour 3 cupfuls with the glass to get 6 oz. Then fill one cup. Pour from one cup into the other until both cups

              are at the same level. Then each has 1 oz. Pour one of them into the glass.

85.   a) $0\times10+0\times9+3\times8+0\times7+0\times6+8\times5+3\times4+6\times3+7\times2=24+40+12+18+14=108$

              $11\overline{)108}=9R9$ $11-9=2$ Yes it checks

              b) $0\times10+7\times9+6\times8+1\times7+7\times6+1\times5+3\times4+2\times3+6\times2=63+48+7+42+5+12+6+12=195$

              $195\div11=17R8$ $11-8=3$ It does not check

87.   A has 2 elements B has 3 elements. $A\times B$ would have $2\bullet3=6$ elements

3.4 Lesson Exercise

LE1   Opener

              a) Use counters

              $2+6=8$

              OO + OOOOOO = OOOOOOOO

              $6+2=8$

              OOOOOO + OO = OOOOOOOO

              b) Yes

              c) Yes

              d) Inductive

LE2   Reasoning

a) No, $2-1 \neq 1-2$

b) Yes

c) No $2 \div 1 \neq 1 \div 2$

LE3   Reasoning

a) Yes

b) Yes

c) Yes

LE4   Reasoning

a) No $(3-2)-1 \neq 3-(2-1)$

b) Yes

c) No $(8 \div 4) \div 2 \neq 8 \div (4 \div 2)$

LE5   Concept

a) No

b) Associative Property

c) $(8 \bullet 3) \bullet x$ or $24x$

LE6   Concept

a) $8 \bullet 7 \bullet (4 \bullet 25)$

b) Commutative and associative properties

LE7   Concept

LE8   Concept

1

LE9   Concept

None

LE10   Concept

None

LE11   Connection

    a) 0 + any number or any number + 0

    b) 1× any number or any number × 1

LE12   Concept

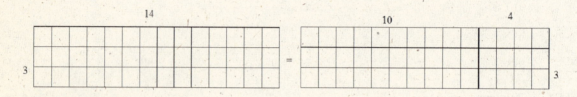

LE13   Skill

    $(7+3) \bullet x$

LE14   Reasoning

    a) Multiplication over subtraction

    b) Yes

    c) Inductive

LE15   Connection

    $24 \times 99 = 24 \times (100 - 1) = (24 \times 100) - (24 \times 1)$

LE16   Opener

Multiplication by zero yields zero. Multiplication by one does not change a number. Commutativity

"eliminates" half of the table, if a child understands that $2 \times 3 = 3 \times 2$ for example.

LE17   Connection

The diagonal of perfect squares, 1, 4, 9, 16 acts a line of reflection that shows the product of two numbers

is not different regardless of how the two numbers are multiplied, the commutative property.

LE18   Skill

a)

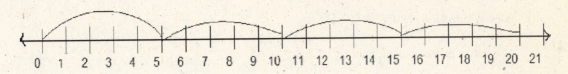

b) $2 \times 6 = 6 + 6$

LE19   Skill

a)

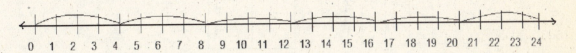

b) $3 \times 7 = 7 + 7 + 7$

LE20   Skill

a) $6 \times 7 = 2 \times (3 \times 7) = 2 \times 21 = 42$

b) $8 \times 7 = 2 \times (4 \times 7) = 2 \times 28 = 56$

LE21   Skill

$1 \times 9 + 6 \times 9 = 9 + 54 = 63$

LE22   Concept

The answer follows the exercise

LE23   Skill

a) $3 \times ? = 15$ The answer is 5

b)

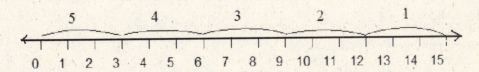

c) Show a total of 15 dots with 3 rows. How many columns are there? 5, so $15 \div 3 = 5$

LE24   Summary

Whole number addition and multiplication is commutative and associative. 0 is the additive identity. 1 is

the multiplicative identity.

SSM Chapter 3   81

3.4 Homework Exercise

1.   The commutative property for addition of whole numbers says $x + y = y + x$, whereas the associative

       property says $(x + y) + z = x + (y + z)$

3.   Associative property of addition

5.

       $(2 \times 4) \times 5$ = $2 \times (4 \times 5)$

7.   Addition and multiplication

9.   a) Commutative property of division

       b) In $8 \div 2$, 2 is the divisor and the result is 4. In $2 \div 8$, 8 is the divisor and the result is ¼.

11.   b

13. $6 \times 9 = 54$ @ $5 per sq ft $50 \times 5 = \$250 + 4 \times 5 = \$20 = \$270$

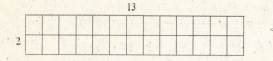   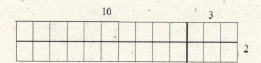

15.

17. $45 \times 98 = 45(100 - 2)$
       $= 4500 - 90$
       $= 4410$

19.   0, identity, addition

21.   a) Associative property of multiplication

       b) Distributive property of multiplication over addition

       c) Commutative property of multiplication

       d) Associative property of addition

23.   Order Property = Commutative

SSM Chapter 3   82

Zero Property = Identity for addition

Grouping Property = Associative

Property of one = Identity for multiplication

25. No, the child is trying to use the distributive property of multiplication over addition

27. a) $10 \times 12 = 120$

   b) $(10 \times 8) + (10 \times 4) = 120$

   c) $10 \times 12 = 10(8+4) = (10 \times 8) + (10 \times 4)$

29. a) True for $A = B$ False for $A \neq B$

   b) True if $B = 1$ or $C = 0$, False if $B \neq 1$ and $C \neq 0$

   c) $A - (B - C) = (A - B) - C \Rightarrow A - B + C = A - B - C$ This is true only if $C = 0$. False for all other cases

   d) True if $A \geq B$ false if $A < B$

   e) True if A is divisible by B. False if A is not divisible by B.

   f) True if $A = 1$ or $C = 0$, false if $A \neq 1$ and $C \neq 0$

   g) One possible answer true if $B = 1$, $A = 2$, and C is any number. False for $A = B = C$

31. The commutative property of addition reduces the number of one-digit addition facts by about half.

33. a)

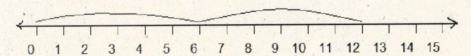

   b) $3 \times 8 = 8 + 8 + 8$

35. a) $6 \times 8 = 2 \times (3 \times 8) = 2 \times 24 = 48$

   b) $8 \times 6 = 2 \times (4 \times 6) = 2 \times 24 = 48$

37. $6 \times 8 = 6(10 - 2)$
   $= 60 - 12$
   $= 48$

39. $8 \times 7 = 56, 7 \times 8 = 56, 56 \div 7 = 8, 56 \div 8 = 7$

41.   a)

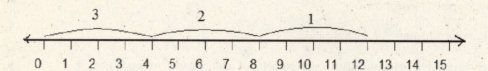

b) Show a total of 12 dots with 4 rows. How many columns are there? 3, so $12 \div 4 = 3$

43.   a) The sum of the digits of the product values is divisible by 9. Each successive product is the previous

product plus 9.

b) Yes

45.   a) It doubles

$$
\begin{array}{cc}
2 & 4 \\
\underline{+3} & \underline{+6} \\
5 & 10
\end{array}
$$

b) $2a + 2b = 2(a+b)$

c) Inductive

d) Deductive

47.   a) Yes

b) Yes, 0

c) Yes

3.5 Lesson Exercise

LE1   Opener

      24
     +32
      56

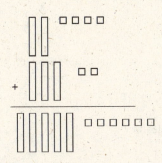

      5 tens 6 units

       24   =   26   Compensation method
      +32   +30
       56   56

       24   =   20   4   50 + 6 = 56 Partial sums algorithm
      +32   +30   +2
       56   50   6

LE2   Connection

      a) Add the ones: 6 ones + 5 ones = 11 ones

      Regroup 11 ones as 1 ten and 1 one

      Add the tens: 1 ten + 2 tens + 3 tens = 6 tens

      The sum is 6 tens 1 one = 61

      b) For student

LE3   Concept

a) 562
    +275
       7
     130
     700
     837

b) 562
    +275
     837

c) Standard algorithm is faster – partial sums algorithm is clearer

LE4   Reasoning

Commutative property of addition

Associative property of addition

Commutative property of addition

LE5   Opener

    35
    −23
    12

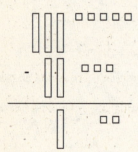

Base – Ten Blocks

    35   5   30   10 + 2 = 12 Partial Differences
    −23   −3   −20
    12   2   10

LE6   Connection

    Step 1   Show 54 as 5 tens and 4 ones.

           Can you take away 8 ones?

    Step 2   Regroup 1 ten as 10 ones.

           Subtract the ones

           14 ones – 8 ones = 6 ones

    Step 3   Subtract the tens

           4 tens – 3 tens = 1 ten

           The difference is 1 ten 6 ones = 16

LE7   Concept

a)
$$
\begin{array}{r}
562 \\
-275 \\
\hline
300 \\
-10 \\
-3 \\
\hline
287
\end{array}
$$

Partial differences algorithm

b) $\overset{4}{\cancel{5}}\,\overset{15}{\cancel{6}}\,{}^{1}2$ Standard algorithm

$$
\begin{array}{r}
+275 \\
\hline
287
\end{array}
$$

c) The standard algorithm is more complicated because it requires re-grouping. The partial algorithm is more efficient because the top digit is less than the bottom digit. The partial algorithm does not require regrouping. The standard algorithm is faster.

LE8   Reasoning

a)
$$
\begin{array}{r}
92 \\
+39 \\
\hline
121
\end{array}
$$

b) The child forgets to carry the group of 10.

c) Use base – 10 blocks to discover the error.

LE9   Reasoning

   a)   921
        −376
        655

   b) The child just subtracts the smaller digit from the larger.

   c) Use base – 10 blocks to discover the error.

LE10   Summary

   Many algorithms can be introduced to explain addition to a student such as the breaking apart algorithm,

   the partial sums algorithm and the expanded algorithm in addition to the standard algorithm. Subtraction

   can be shown with the partial differences algorithm.

3.5 Homework Exercises

1.   a) Correct, the child understands place value and breaking apart numbers

      b) Incorrect, use base – 10 blocks to show the child how to regroup and trade 14 units for 1 long and 4

      units.

      c) Correct, the child understands place value and adding on.

      d) Correct, the child understands compensation with addends and adding by tens.

3.   Combine 2 units and 6 units to get 8 units

      Combine 8 longs and 3 longs to get 1 flat and 1 long

      Combine 1 flat and 3 flats to get 4 flats

      No, add 1 flat + 4 flats + 1 long + 8 units = 5 flats 1 long 8 units 518

5.   a)   357
          +529
            16
            70
           800
           886

      b) 3 5 7
          +529
           886

      c) The standard algorithm is faster. The partial sums algorithm helps the child to understand regrouping.

7. Associative property, commutative property of addition

Associative property, commutative property of addition

Associative property

9. a) No, only one number has to be greater than 300. For example $1000 + 50 + 150 = 1200$

b) Yes the two largest numbers would

11. a) Subtraction as adding on

b) Place value and breaking numbers apart

c) Subtraction as adding on

13. 7 hundreds 12 tens 7 ones

15.   336
      -182

Take 2 units from the 6 units to leave 4 units. You cannot take 8 longs from 3 longs –regroup 3 flats as 2 flats and 10 longs. Now you can take 8 longs from the 13 longs and finally take 1 flat from 2 flats. Answer – 1 flat 5 longs 4 units.

17. a)   814
       -391
       500
       -80
       $+3$
       423

b)   8 ^{1}14
     -391
     423

c) Regrouping is easier to understand in the partial differences algorithm. The standard algorithm is faster.

19. They are less likely to forget to add it.

21. a) $50 + 20$

b) $50 - 20$

23. c), b), a) In order of difficulty

25. a) 87
         $\underline{+8}$
         175

    b) Adding the units digit to both the tens and the units place

    c) Adding the number of ones in the second addend to the number of tens in the first addend

27. a) 93
         $\underline{+28}$
          75

    b) Subtracting from left to right and incorrect borrowing

    c) Use base – ten blocks

29. a) $29+1$ b) $137+3$
        $30+30$ $140+60$
        $\underline{60+2}$ $\underline{200+12}$
        $62+33$ $212+75$
        $62-29=33$ $212-137=75$

31. a) Could use partial sums and counting on. 35, 45, 55, 64

         35
        $\underline{+29}$
         50
        $\underline{14}$
         64

    b) Could use partial differences and counting back. 41, 31, 21, 15

         41
        $\underline{-26}$
         20
        $\underline{-5}$
         15

33. a) $\overset{2}{3}8$
         $+\cancel{9}\,_4\cancel{7}\,_5$
         $\overset{1}{2}4\cancel{6}\,_1$
          381

    b) You don't have to remember large sums in your head

35.   a) $86 - 29 = 57$

       b) $87 - 30 = 57$ Yes

       c) $97 - 40 = 57$ Yes

37.   a) $\begin{array}{r} 72 + 3 \\ -47 + 3 \end{array}$ $\begin{array}{r} 75 \\ -50 \\ \hline 25 \end{array}$

       b) $\begin{array}{r} 821 + 4 = 825 \\ -376 + 4 = -380 \\ \hline 445 \end{array}$

39.   For student

3.6 Lesson Exercises

LE1   Opener

       Express 34 as 3 longs and 4 units. Now, double the 3 longs to get 6 longs, and double the 4 units to get 8

       units. Your answer is 68.

LE2   Connection

       Step 1   Shows 3 groups of 26

       Step 2   Combine the ones. You have 18 ones. Trade 10 ones for 1 ten leaving 8 ones

       Step 3   Combine the tens. You have 6 tens plus 1 ten for a total of 7 tens. The product is 7 tens 8 ones or

                 78. So $26 \times 3 = 78$

LE3   Reasoning

       Distributive Property

       Commutative property of addition

LE4   Opener

         26   Partial products algorithm
        ×43
         18
         60
        240
        800
       1118

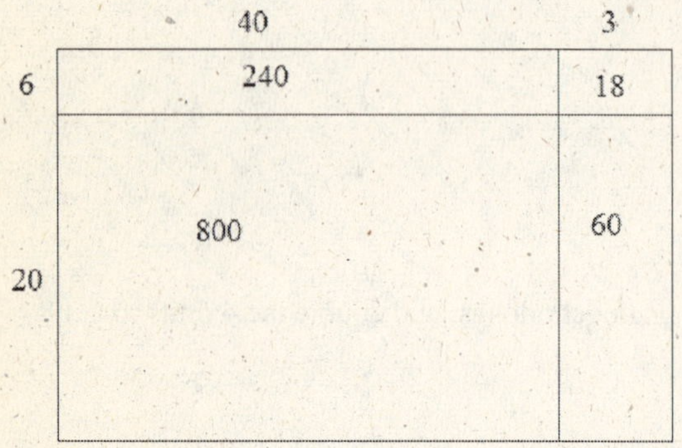

	40	3
6	240	18
20	800	60

=1118.

LE5   Connection

a) , b)

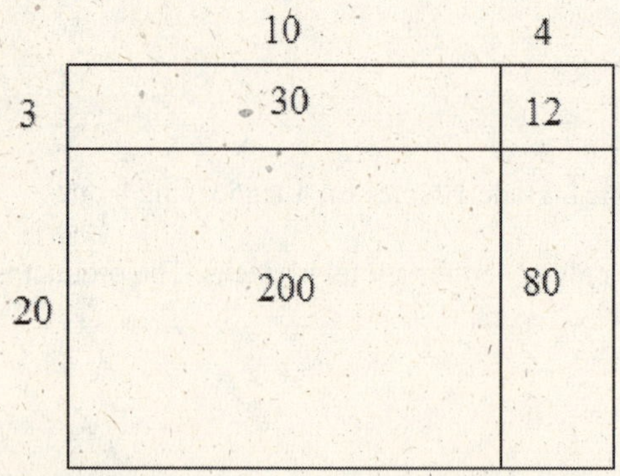

	10	4
3	30	12
20	200	80

c) $30 + 12 + 200 + 80 = 322$

LE6   Concept

    a)   83 Partial products algorithm

$$
\begin{array}{r}
83 \\
\times 47 \\
\hline
21 \\
560 \\
120 \\
3200 \\
\hline
3901 \\
\end{array}
$$

    b)   83

$$
\begin{array}{r}
83 \\
\times 47 \\
\hline
581 \\
332 \\
\hline
3901 \\
\end{array}
$$

c) It's easier to see how place value works in the partial products algorithm. The standard algorithm is

shorter and faster.

LE7   Opener

With base – 10 blocks show 4 groups of

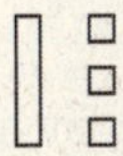

(13). There will be 1 left over. Therefore the answer is 13 R 1.

LE8   Connection

Show 7 tens and 1 unit. Divide the tens into 3 equal groups. Put 2 tens in each group. There is 1 ten left

over. Regroup the ten as ten ones. This makes 10 ones and 1 one or 11 ones. Try to divide the 11 ones into

3 equal groups. Put 3 ones in each group there are 2 ones left over. The quotient is 23 R 2.

LE9   Concept

a) $6\overline{)394}$

$$\underline{300}$$
$$94$$
$$\underline{60}$$
$$34$$
$$\underline{30}$$
$$4$$

b) $\overset{65}{6\overline{)394}}$ Which results in 65 R 4

$$\underline{-36}$$
$$34$$
$$\underline{30}$$
$$4$$

c) Repeated subtraction is easier to understand but longer. The standard algorithm is quicker but not as

clear.

LE10   Reasoning

a)
$$\begin{array}{ccc}
36 & 42 & 72 \\
\underline{\times 8} & \underline{\times 6} & \underline{\times 9} \\
568 & 302 & 728
\end{array}$$

b) Child multiplies the sum of the number in the tens position and the carried number.

c) Use base – 10 blocks to model the problem.

LE11   Reasoning

a) $\overset{221}{3\overline{)782}}$

b) The child always divides the smaller number into the larger.

c) Let's use base – 10 blocks to model the division problem.

LE12   Summary

Multiplication can be shown with the partial products algorithm to promote understanding. Multiplication can also be shown with an area model as well as the standard algorithm. Division can be shown as repeated subtraction.

3.6 Homework Exercises

1.   a) Breaking apart numbers and the distributive property of multiplication over addition

b) Same as a)

c) Breaking apart numbers and the distributive property of multiplication over subtraction

3.

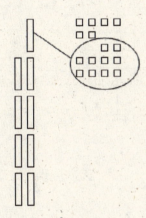

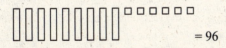 $= 96$

5.   Partial Products Algorithm

39×48

$= (30+9) \times (40+8)$ Expanded Notation

$= [(30+9) \times 40] + [(30+9) \times 8]$ Distributive Property of Multiplication over Addition

$= [(30 \times 40) + (9 \times 40)] + [(30 \times 8) + (9 \times 8)]$ Distributive Property

$= [(30 \times 8) + (9 \times 8)] + [(30 \times 40) + (9 \times 40)]$ Commutative Property of Addition

$= [(9 \times 8) + (30 \times 8)] + [(9 \times 40) + (30 \times 40)]$ Commutative Property of Addition

7.

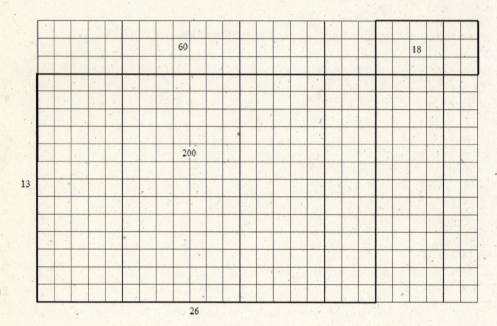

```
    13
   ×26
    18
    60
    60
   200
   338
```

9.

a) Partial Products Algorithm

```
   49
  ×62
   18
   80
  540
 2400
 3038
```

b) Standard Algorithm

```
   49
  ×62
   98
  294
 3038
```

c) The partial products algorithm allows the student to see the use of the distributive property. The standard algorithm is much faster.

11. 2×40 , 50×40

13.   a) Correct, child understands repeated subtraction

b) Incorrect, use base – 10 blocks to model 96 and have the child see how many groups of 8 the can form

15.

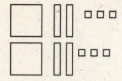

Divide each into 2 equal parts

Standard algorithm

$$\frac{123}{2\overline{)246}}$$

17.   a) $217 \div 4$ repeated subtraction $4\overline{)217}$ Ans 54 R 1

$$
\begin{array}{rl}
200 & 50 \times 4 \\
17 & \\
-16 & 4 \times 4 \\
\hline
1 &
\end{array}
$$

b) $217 \div 4$ Standard algorithm

$$
\begin{array}{r}
54R1 \\
4\overline{)\,217} \\
20 \\
17 \\
\underline{16} \\
1
\end{array}
$$

19.   d), b), a), c) In order of difficulty

21.   a) 72   b) Child forgot to carry the group of 10.   c) Go over multiplication and regrouping
$$
\begin{array}{r}
\times 9 \\
\hline
638
\end{array}
$$

23.   a) $8\overline{)512}$ with 46 above, 48 below, 32

        b) Placing the digits in the quotient from right to left

        c) Discuss place value and the placement of digits in the quotient

25.   a) The child has only multiplied two of the partial products

        b)

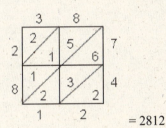

        Child is leaving out 10×6 and 20×5

27.   a)

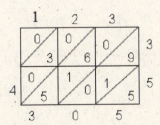

                              = 2812

        b)

                              = 4305

        c) It is harder to understand the underlying concepts of this algorithm

29.   a) 56×7 - use base – 10 blocks use lattice multiplication

        b) 41×26 - use lattice multiplication and partial products

        c) $86\div20$ use repeated subtraction, use base – ten blocks

31. $8659 \div 23$ quotient $= 372$ Remainder $.56522 \times 23 = 13$

3.7 Lesson Exercises

LE1   Opener

$$\$46 + \$79 \text{ add } 40 + 70 = 110$$
$$6 + 9 = \underline{15}$$
$$125$$

LE2   Reasoning

a) Abdul and Callista used the breaking apart strategy based on place value.

b) Ramon used compensation.

c)

Abdul: $50 + 30 = 80$ and $8 + 6 = 14$ then $80 + 14 = 94$

Callista: $58 + 30 = 88$ and $88 + 6 = 94$

Ramon: $58 + 36 = 54 + 40 = 94$

d) Commutative property

e) You could start at 58 and count on 36 to end at 94.

LE3   Opener

$93 - 33 = 60$ and $60 - 5 = 55$

LE4   Reasoning

Abdul: $62 - 20 = 42$ and $42 - 9 = 33$

Callista: $29 + 1 = 30$ then $30 + 32 = 62$ the answer is $1 + 32 = 33$

Ramon: $62 - 29 = 63 - 30 = 33$

LE5   Opener

a) 476

b) 24000

LE6   Reasoning

$13 \times 28 = 13 \times 30 - 13 \times 2 = 390 - 26 = 364$

LE7   Opener

Frequently estimation is used in the grocery store.

LE8 Opener

a) It is not possible to find the exact answer.

b) You use an over estimate to be safe.

c) You cannot know the exact time in advance.

LE9 Skill

Estimate 93×18 by rounding to 90 and 20, $90 \times 20 = 1800$ calories.

LE10 Skill

Round the problem to $59,000 - 3,000 = 56,000$ tickets

LE11 Reasoning

a) Too high (Hint try $29 - 11$)

LE12 Opener

$712 \div 32$, make an easier problem $700 \div 35 = 20$

LE13 Skill

a) $19847 \div 24 \approx 20,000 \div 20 = 1,000$

b) $198 \div 24 \approx 200 \div 25 = 8$

LE14 Reasoning

a) Too high (Hint try $29 \div 11$)

LE15 Skill

a) $36 + 71$, $52 + 51$, 98 an d 96 are each about 100. The sum is about 400.

b) Commutative and associate properties of addition

LE16 Skill

Add $3 + 4 + 2$, thousand $= \$9000$ then $793 + 391 + 807$ is about another $2000. Compute

$9000 + \$2000 = \$11,000$

LE17 Reasoning

For the student to play this game.

LE18 Summary

a) A method of mental computation could be the use of the distributive property.

b) $13 \times 28 = 13 \times 30 - 13 \times 2 = 390 - 26 = 364$

3.7 Homework Exercises

1. a) $39 + 97$ can be expressed as : $30 + 90 + 9 + 7 = 120 + 16$ or $36 + 3 + 97 = 36 + 100 = 136$ or
$$= 136$$

$39 + 90 = 129 + 7 = 136$

b) Commutative property of addition

3. $134 - 58$ can be expressed as:　　$134 - 50 = 84$ and $84 - 8 = 76$,

$134 - 58 = 136 - 60 = 76$,

$130 - 50 = 80$ and $8 - 4 = 4$ and $80 - 4 = 76$

5. a) $236 - 1 = 235$

$$\underline{+89 + 1 \quad +90}$$
$$325$$

b) $82 + 2 = 84$

$$\underline{-58 - 2 \quad -60}$$
$$24$$

7. $39¢ \times 6 =$　　a)　　$30 \times 6 + 9 \times 6 = 180 + 54 = \$2.34$

b)　　$40 \times 6 - 6 = 240 - 6 = \$2.34$

c)　　Distributive property

9. $500^3 \rightarrow 5^3 = 125$ and $100^3 = 1000000 \quad 125 \times 10^6 = 125,000,000$

11. a) You must estimate when you predict the future

b) You don't know exactly how much each person weighs

13. a) $700 + 500 + 700 = 1900$ Not reasonable

b) $30 \times 50 = 1500$ Reasonable

15. $5692 \approx 6000$
$8091 \approx 8000$
$3721 \approx \underline{4000}$
$ 18,000$

17. a) Change $437 \div 55$ to compatible numbers like $40 \div 50 = 8$ hours

b) Equal measures

19. Too high

21. $469 \div 74 \approx 420 \div 70 = 6$

23. $59 + 42 \approx 100$ and $97 \approx 100 : 100 + 100 = 200 + 32 \approx 230$

25.   a) $8 \times 5 = 40$ and $22 \times 40 = 880$ Commutative property of multiplication

        b) $(2 + 78) + 43 = 80 + 43 = 123$ Associative property

27.   Front End

$$
\begin{array}{r|l}
 & 3{,}462{,}871 \\
 & \quad\quad\quad \}\text{Adjust} = \\
 & 830{,}212 \quad 26{,}500{,}000 \\
2 & 1{,}172{,}806 \\
\hline
2 & 0{,}000{,}000 \\
 & 6{,}500{,}000 \\
\hline
2 & 6{,}500{,}000
\end{array}
$$

        Rounding $3 + 1 + 21 = 25{,}000{,}000$

29.   a) $32 \times 48 \approx 30 \times 40 = 1200$ light bulbs

        b) $32 \times 48 \approx 30 \times 50 = 1500$ light bulbs

        c) Rounding

31.   a) $4872 - 3194 \approx 5000 + 3000 = 2000$

        b) $3279 \div 65 \approx 3000 \div 60 = 50$

33.   a) Calculator – takes too long otherwise

        b) Mental computation (add 750 and 250 first)

        c) Paper and pencil or calculator – whichever is handier

35.   a) $50 \times 30 = 1500$ and $60 \times 40 = 2400$ Between 1500 and 2400

        b) $8000 + 2000 = 10000$ and $9000 + 3000 = 12000$ Between 10,000 and 12,000

        c) $300 \div 10 = 30$ and $400 \div 10 = 40$ Between 30 and 40

37.   Yes Multiplying by 5 is the same as dividing by 2 and then multiplying by 10. (Adding a zero at the end)

39.   a) Taking ten 74's, doubling to get twenty 74's then adding two 74's to end up with twenty – two 74's

        b) $31 \times 33 = 31 \times 10 = 310$ then $3 \times 310 = 930$ then $930 + 93 = 1023$

41.   a) $5 \bullet 800 = 4000$

        b) $4 \bullet 40,000 = 160,000$

43.   a) 21

        b) 16

        c) 38

45.   a) 631×542

        b) Same problem but use 123789 as the digits $931 \times 872 = 811832$

47.   a) 14

        b) 177

        c) 7

49.   a) 8

        b) 28

        c) 36

        d) 26

        e) 17

51.   Left for student

53.   For the student

3.8 Lesson Exercises

LE1   Opener

      $A = 1, B = 2, C = 3, D = 4, E = 0$

      This represents a base – five system

$0 = E$	$5 = AE$
$1 = A$	$6 = AA$
$2 = B$	$10 = B0$
$3 = C$	$30 = AAE$
$4 = D$	$50 = BEE$
	$100 = DEE$

LE2   Concept

      0, 1, 2, 3, 4

LE3   Skill

$$(3\times8)+(2\times1)=24+2=26$$

LE4   Skill

$$426_{eight}=(4\times8^2)+(2\times8^1)+(6\times8^0)$$
$$=256+16+6$$
$$=278 \text{ in base ten}$$

LE5   Skill

$$\frac{2}{5^3}\frac{1}{5^2}\frac{3}{5^1}\frac{4}{5^0}_{five}=(2\times5^3)+(1\times5^2)+(3\times5^1)+(4\times5^0)$$
$$=250+25+15+4$$
$$=294$$

LE6   Opener

a) 12 cents = 2 nickels, 2 pennies

b) 83 cents = 3 quarters, 1 nickel, 3 pennies

LE7   Skill

a) 302 $8^0=1$

$8^1=8$

$8^2=64$

$8^3=512$

$$=456_{eight}$$

$$64\overline{)302}^{\,4} \ (8^2)$$
$$\underline{256}$$
$$46$$

$$8\overline{)46}^{\,4} \ (8^1)$$
$$\underline{40}$$
$$6 \ (8^0)$$

b) 302 $5^0=1$

$5^1=5$

$5^2=25$

$5^3=125$

$$=222_{five}$$

$$125\overline{)302}^{\,2} \ (5^3)$$
$$\underline{250}$$
$$52$$

$$25\overline{)52}^{\,2} \ (5^1)$$
$$\underline{50}$$
$$2 \ (5^0)$$

LE8   Skill

a) 0, 1, 2, 3, 4, 10, 11, 12, 13, 14, 20

b)

+	0	1	2	3	4
0	0	1	2	3	4
1	1	2	3	4	10
2	2	3	4	10	11
3	3	4	10	11	12
4	4	10	11	12	13

LE9   Connection

a) 24_{five} Partial sums

$+14_{five}$

13

30

43_{five}

b) $24_{five} + 14_{five}$ Add the 4 units to the 4 units, that gives 8 units. Trade 5 units for 1 long, 3 units are left.

Add 2 longs and 1 long and 1 long to get 4 longs: Ans: 4 longs 3 units Base – 5 blocks

LE10   Connection

a) $3\cancel{2}_{five}$

$+14_{five}$

3

10

13_{five}

b) Trade 1 long for 5 units. Add 5 units to your 2 units, resulting in 2 longs 7 units. Now subtract 4 units

from 7 units. You are left with 3 units. 1 long from 2 longs leaves 1 long. Ans: 1 long 3 units.

LE11   Skill

	0	1	2	3	4
0	0	0	0	0	0
1	0	1	2	3	4
2	0	2	4	11	13
3	0	3	11	13	22
4	0	4	13	22	31

LE12   Connection

a) 44_{five} Partial products algorithm

$\times 22_{five}$

13

130

130

1300

2123_{five}

b) 44_{five} Standard Algorithm

$\times 22_{five}$

143

143

2123_{five}

LE13   Connection

a) $3_{five}\overline{)1343_{five}}$ 200×3 Ans: 244 R 3

$$\underline{1100}$$
$$243 \qquad 20\times3$$
$$\underline{110}$$
$$133 \qquad 20\times3$$
$$\underline{110}$$
$$23 \qquad 4\times3$$
$$\underline{22}$$
$$1$$

Repeated subtraction

b) $3_{five}\overline{)1343_{five}}^{\;244R1}$

$$\underline{-11}$$
$$24$$
$$\underline{22}$$
$$23$$
$$\underline{22}$$
$$1$$

LE14   Summary

Studying place value and the algorithms is less familiar number systems deepens your understanding of place value and the algorithms. Base − ten and base − five each have a units place. The base changes as you move away from the units place.

3.8 Homework Exercises

1. a) 1 eight and 4 ones

    b) 2 fives and 2 ones

3. $0_{three}1_{three}2_{three}10_{three}11_{three}12_{three}2_{three}21_{three}22_{three}100_{three}101_{three}102_{three}110_{three}$

5. a) $75_{eight} = 7\times8+5\times1 = 61$

    b) $423_{five} = 4\times25+2\times5+3\times1 = 113$

    c) $213_{eight} = 2\times64+1\times8+3\times1 = 139$

7.   a) $46 = 5 \times 8 + 6 \times 1 = 56_{eight}$

        b) $26 = 1 \times 25 + 1 \times 1 = 101_{five}$

        c) $324 = 2 \times 125 + 2 \times 25 + 4 \times 5 + 4 \times 1 = 2244_{five}$

9.   The last digit is 0

11. 1314_{five}

13.   a) 13_{five}

$$\begin{array}{r} +22_{five} \\ \hline 10 \\ 30 \\ \hline 40_{five} \end{array}$$

        b)

$$\begin{array}{r} 13_{five} \\ +22_{five} \\ \hline 40_{five} \end{array}$$

15.   a) 43_{five}

-24_{five}

Partial Differences

You cannot take 4 ones from 3 ones, so trade 4 fives for 3 fives and 5 ones. Now you have 3 ones and 5

ones or 8 ones. Subtract 4 ones from 8 ones and you get 4 ones. Finally subtract 2 longs from 3 longs and

you get 1 long.

b)

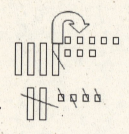

$\overset{3\ \ 5}{\cancel{4}\,3}_{five}$ Standard algorithm

-24_{five}

14_{five}

17.   a) 23_{five}

$\underline{+34_{five}}$

112_{five}

      b) 324_{five}

$\underline{+132_{five}}$

1022_{five}

      c) 432_{five}

$\underline{+233_{five}}$

1220_{five}

19.   a) 23_{five}

$\underline{\times 14_{five}}$

22

130

30

$\underline{200}$

432_{five}

      b) 23_{five}

$\underline{\times 14_{five}}$

202

$\underline{23}$

432

21.   a)

$$3_{five} \overline{)1\overset{5}{2}1_{five}} \quad \text{with } 22 \text{ above}$$

$\underline{22}$ 4×3

44

$\underline{22}$ 4×3

22

$\underline{22}$ 4×3

0

   Ans = 22_{five}

     b) $3_{five}\overline{)121_{five}}$ Ans 22_{five} (quotient 22)

$\underline{11}$

11

$\underline{11}$

23.   a) 34_{five}

           $\times 23_{five}$

           212

          123

       1442_{five}

    b) 412_{five}

          $\times 321_{five}$

          412

        1324

       2241

     243302_{five}

25.

+	0	1	2	3	4	5	6	7
0	0	1	2	3	4	5	6	7
1	1	2	3	4	5	6	7	10
2	2	3	4	5	6	7	10	11
3	3	4	5	6	7	10	11	12
4	4	5	6	7	10	11	12	13
5	5	6	7	10	11	12	13	14
6	6	7	10	11	12	13	14	15
7	7	10	11	12	13	14	15	16

×	0	1	2	3	4	5	6	7
0	0	0	0	0	0	0	0	0
1	0	1	2	3	4	5	6	7
2	0	2	4	6	10	12	14	16
3	0	3	6	11	14	17	22	25
4	0	4	11	14	20	24	30	34
5	0	5	12	17	24	31	36	43
6	0	6	14	22	30	36	44	52
7	0	7	16	25	34	43	52	61

27.   a)
$$
\begin{array}{r}
25_{eight} \\
4_{eight} \overline{)124_{eight}} \\
\underline{10} \\
24 \\
24
\end{array}
$$

b)
$$
\begin{array}{r}
244_{eight}\ R2_{eight} \\
3_{eight} \overline{)756_{eight}} \\
\underline{6} \\
15 \\
\underline{14} \\
16 \\
\underline{14} \\
2
\end{array}
$$

29.   a) $1101_{two} = (1 \times 2^3) + (1 \times 2^2) + (1 \times 1) = 13_{two}$

b) $17 = 10001_{two}$

c) $1101_{two} + 111_{two} = 10010_{two}$

31. $555 eight = (5 \times 8^2) + (5 \times 8) + (5 \times 1) = 320 + 40 + 5$
$$= 365$$

a) $365 \rightarrow$ base $2 = 101101101_{two}$

b) Convert each base – eight digit to an equivalent three digit base – two numeral and place the resulting numerals in the same position as the original base – eight digits.

c) 110100010_{two}

33.   a) $E6 = (11 \times 12) + (6 \times 1) = 132 + 6 = 138$

b) $80 \rightarrow$ base 12, $80 = 68_{twelve}$

35.   a) 1, 2, 3, 4, … 15 pounds

        b) 1, 2, 4, and 8 16 weights are needed

        c) Same problem with 1 to 100 lbs. Need weights 1, 2, 4, 8, 16, 32, 64

37.   a) A = 1 B = 4 C = 6 0 = 0 There is no place value in this example

        b) A, AA, AAA, B, BA, BAA, BAAA, BB, BBA, BBAA

        c) CCCAA

39.   a) Yes

        b) All other bases

41.   For the student

Chapter Three Review Exercises

1.   a) 37

        b) 2

        c) 6

3. $125,000 - 134,499$ (inclusive)

5.   a) Groups

        b) Compare

7.   A equal groups problem asks how many times you must repeat the amount of the divisor to obtain the total amount. A partition a group problem asks you to divide up the total into the number of equal sized groups specified by the divisor.

9.   a) Measures

        b) Subtraction, Take away and Division, Equal

11.   Bob is going to Florida for 5 days and Connecticut for 6 days. How long will he be gone?

13.   I had 8 marbles, now I only have 3. How many are missing?

15. x = amount of money to start

$$x - \underset{\text{groceries}}{30} - \underset{\text{book}}{\frac{1}{2}(x-30)} = \underset{\text{amount left}}{8}$$

$$x - 30 - \frac{1}{2}x - 15 = 8$$

$$\frac{1}{2}x - 15 = 8$$

$$\frac{1}{2}x = 23$$

$$x = \$46$$

17. a) No

   b) $(10 \div 5) \div 2 \overset{?}{=} 10 \div (5 \div 2)$

   $2 \div 2 \overset{?}{=} 10 \div 2.5$

   $1 \neq 4$

19. a) $8 \times 40 = 320$

   b) Distributive property of multiplication over addition

21.
$$\begin{array}{r} 326 \\ +293 \\ \hline \end{array}$$

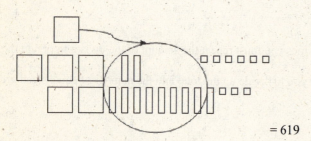

$= 619$

23. a) Subtraction as adding on

   b) Compensation

   c) Marc: Count up 3 to 40 and 21 more to 61. The answer is $3 + 21 = 24$

   Julia: I changed $61 - 37$ to $64 - 40 = 24$

25.   26
       ×3

Take 2 longs three times = 6 longs

Take 6 ones three times = 18 ones = 1 long and 8 ones

6 longs + 1 long + 8 ones = 7 longs and 8 ones

Standard

    ¹26
    ×3
    78

27.   a) 89
           74
          1513

b) Child is forgetting to carry

c) Use base – 10 blocks

29.   a)   ⁵48
           ×57
           336
           250
          2836

b) Forgetting to add in the correct carried number

c) Use partial products

31.

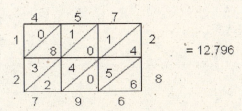

= 12,796

33.   a) Change to compatible $2400 \div 40 = 60 \; gallons$

b) Measures

c) Equal

35. $100 = 244_{six}$

37.   a)

$$6_{seven}\overline{)\,{}^{0}1\,{}^{7}324_{seven}}$$

$$
\begin{array}{ll}
\underline{600} & 100\times6 \\
424 & \\
\underline{150} & 20\times6 \\
244 & \\
\underline{150} & 20\times6 \\
64 & \\
\underline{60} & \\
4 &
\end{array}
$$

Ans 150_{seven} R 4

b)

$$150R4$$
$$6_{seven}\overline{)\,{}^{0}1\,{}^{7}324_{seven}}$$

$$
\begin{array}{l}
\underline{-6} \\
42 \\
\underline{42} \\
4
\end{array}
$$

Chapter Four

4.1 Lesson Exercises

LE1   Opener

       a) 1 and 2 (the factors)

       b) Any multiple of x and y

LE2   Concept

       a) $21 \div 3 = 7$

       b) $3 \bullet 7 = 21$

LE3   Concept

       a) 2 is a factor of 40 because $2 \bullet 20 = 40$

       b) $2 \bullet N = R$

       c) Deductive

LE4   Concept

       a) 1, 2, 3, 4, 6, 12

       b) False

       c)

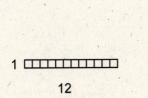

LE5   Reasoning

       863,1<u>02</u> (The last digit is 2, 4, 6, 8 or 0. The other digit can be any number)

LE6   Reasoning

        True

LE7   Concept

        A times some whole number equals B

LE8   Reasoning

        Show $A|(B+C)$

        1. $A|B$ and $A|C$

        2. $A \bullet D = B$ and $A \bullet E = C$, in which D and E are whole numbers

        3. $A \bullet D + A \bullet E = B + C$

        4. $A \bullet (D + E) = B + C$ in which $D + E$ is a whole number

        5. $A|B + C$

LE9   Reasoning

        True

        1. $A|B$

        2. $AD = B$ in which D is a whole numbers

        3. $A(DC) = BC$ in which DC is a whole number

        4. $A|BC$

LE10   Reasoning

        False $2|4 \bullet 3$, $2|4$ but 2 does not divide 3

LE11   Reasoning

        $5|4200$

LE12   Reasoning

        $A|3C$

LE13   Communication

        $6|12 \bullet 37$: $6|12$ so 6 is a factor of 12 times any whole number such as $12 \bullet 37$

LE14   Skill

For the student to play game.

LE15   Summary

A is a factor of B if and only if there is a whole number C such that $A \bullet C = B$. A perfect number equals the sum of all its factors that are less than itself. By the divisibility of the sum theorem, for any whole numbers A, B, and C, with $A \neq 0$, if $A|B$ and $A|C$ then $A|(B+C)$. By the divisibility of a difference theorem, for any whole numbers A, B, and C, with $A \neq 0$ and $C > B$, if $A|B$ and $A|C$ then $A|(C-B)$. By the divisibility of a product theorem with A, B, and C whole numbers and $A \neq 0$ if $A|B$ then $A|BC$.

4.1 Homework Exercises

1.   a) 1, 2, 4, 7, 14, 28

b) $6 \bullet C \neq 28$ for $C \in W$

3.   True

5.   a) 1 by 20, 2 by 10, 4 by 5

b) Factoring 20

7.   False; $2|4$ but $4 \nmid 2$

9.   3, B, A, B

11.   Divisibility of a product theorem

13. $8|(C+16)$

15.   False; $5|5$ but 10 does not divide 5

17.   True; $A = 2$, $B = 4$, $C = 3, D = 6$

19.   True;

1. $A|B$, $B|C$ and $A \neq 0$

2. $A \bullet K = B$ and $B \bullet L = C$ $K, L \in W$

3. $A \bullet K \bullet L = C$ and KL is a whole number

4. $A|C$

21.   True

   1. C and D are even numbers

   2. $C = 2w$ $D = 2x$ where w and x are whole numbers

   3. $CD = 2w \bullet 2x$

   4. $CD = 2(2wx)$

   5. CD is an even number

23. $x|x$ and $x|xy^2 \rightarrow x|(xy^2 \bullet x)$ or $x|x^2y^2$

25.   a) 20 $1, 2, 4, 5, 10, \cancel{20}$ $1 + 2 + 4 + 5 + 10 = 22$ abundant

   b) 28 $1, 2, 4, 7, 14, \cancel{28}$ $1 + 2 + 4 + 7 + 14 = 28$ perfect

   c) 38 $12, 19, \cancel{38}$ $1 + 2 + 19 = 22$ deficient

27.   a) $2 \bullet 3 \bullet 4 \bullet 5 + 1 = 121$

   b) $3 \bullet 4 \bullet 5 \bullet 6 + 1 = 360 + 1 = 361$

   c) The result is always a perfect square number

   d) $n(n+1)(n+2)(n+3) = n^4 + 6n^3 + 11n^2 + 6n + 1$
   $$= (n^2 + 3n + 1)^2$$

4.2 Lesson Exercises

LE1   Opener

   A number is divisible by

   2 – if two divides the last digit. i.e. – 0, 2, 4, 6, 8

   3 – if three divides the sum of the digits

   4 – if four divides the last two digits

   5 – if five divides the last digit. i.e. 0 or 5

   6 – if both 2 and 3 divide the number

   8 – if the last digits of the number is divisible by 8

   9 – if nine divides the sum of the digits

   10 – if ten divides the last digit. i.e. 0

$2|200000022, 3|200000022, 6|200000022$

LE2 Concept

  0, 8, 16, 24, 32, …

LE3 Concept

  a) factor

  b) multiple

  c) multiple

  d) multiples

LE4 Reasoning

  a) its ones digit is divisible by 2

  b) its ones digits is 0 or 5

  c) its ones digit is 0

LE5 Skill

  a) 2

  b) 2, 5, 10

  c) 5

  d) Deductive

LE6 Reasoning

  The last digit is 2, 4, 6, or 8. The other digit can be any number.

LE7 Opener

  a) If the sum of the digits in the number are divisible by 3 then the number itself is divisible by 3, if the sum

  of the digits in the number are divisible by 9 then the number itself is divisible by 9.

  b) Any whole number that is divisible by 9 is also divisible by 3.

LE8 Skill

  a) Neither

  b) 3, 9

LE9   Reasoning

    a) The answer will be correct if the sum of the digits is divisible by 9.

    b) Sum the digits and determine two other digits that will make the sum divisible by 9.

    c) Any two digits with a sum of 8.

    d) No

LE10   Reasoning

    a) 224, 228

    b) 1224, 1228

    c) The last two digits are divisible by 4.

    d) If the last two digits are divisible by 4 then the number is divisible by 4.

LE11   Reasoning

    a) The numbers are all divisible by 2 and 3

    b) A whole number is divisible by 6 if and only if the number is divisible by both 2 and 3

LE12   Skill

    8172 divisible by: 2, 3, 4, 6, 9

    403,155 divisible by: 3, 5, 9

    800,002 divisible by: 2

    68,710 divisible by: 2, 5, 10

LE13   Reasoning

    False $2 \mid 4$ and $4 \mid 4$ but $8 \nmid 4$

LE14   Reasoning

    True   7 is an example

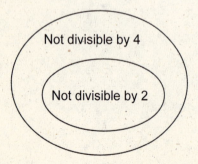

LE15   Opener

a) 3000; 600; 40

b) Yes

c) Divisibility of a Sum

LE16   Reasoning

1. $\underline{AB0}$ and $\underline{AB5}$ are three digit numbers

2. $\underline{AB5} = (A \bullet 100) + (B \bullet 10) + 5$ and
        $\underline{AB0} = (A \bullet 100) + (B \bullet 10)$

3. $(A \bullet 100)$ and $(B \bullet 10)$ and 5 are all divisible by 5

4. $(A \bullet 100) + (B \bullet 10) + 5$ and $(A \bullet 100) + (B \bullet 10)$ are divisible by 5

5. $\underline{AB0}$ and $\underline{AB5}$ are divisible by 5

LE17   Reasoning

a) $\underline{ABCD}$ is a four digit number with $A + B + C + D$ divisible by 3

b) $(A \bullet 1000) + (B \bullet 100) + (C \bullet 10) + D$

c) $A + B + C + D$

d) Divisible by 3

e) Divisible by 3

LE18   Communication

To check for divisibility by 3 and 9, you must find the sum of the digits. If the sum is divisible by 3 and 9 then the number is divisible by 3 and 9.

LE19   Summary

For divisibility by 2, 5, and 10, you check the last digit for divisibility by those numbers. For divisibility by 3 and 9, the sum of the digits must be divisible by 3 and 9 respectively. For a number to be divisible by 6 it must be divisible by 2 and 3.

4.2 Homework Exercises

1. $0, 7, 14, 21, \ldots$

3.   a) factor

         b) factor

         c) multiple

         d) factor

5.   30, 15, 5, 45, and 3 since there would be at least one number in each region.

7.   No. "A number is divisible by 4 if and only if the last two digits represent a number that is divisible by 4."

9.   a) 7533 $2:$ yes $2 \nmid 3$ $3:$ yes $3 \mid 18$ $4:$ no $4 \nmid 33$ $5:$ no $5 \nmid 3$
               $6:$ no $2 \nmid n$ $9:$ yes $9 \mid 18$ $10:$ no $10 \nmid 3$

         b) 1344 $2:$ yes $2 \mid 4$ $3:$ yes $3 \mid 12$ $4:$ yes $4 \mid 44$ $5:$ no $5 \nmid 4$
               $6:$ yes 2 and $3 \mid 1344$ $9:$ no $9 \nmid 12$ $10:$ no $10 \nmid 4$

         c) 410,333 $2:$ yes $2 \mid 0$ $3:$ no $3 \nmid 11$ $4:$ no $4 \nmid 30$ $5:$ yes $5 \mid 0$
               $6:$ yes $3 \nmid n$ $9:$ no $9 \nmid 11$ $10:$ the last digit is 0.

11.   no 9 does not divide 219

13.   a) 102, 108, 114, 120, 126

         b) The numbers are all even and the sum of the digits is divisible by 3

         c) For a number to be divisible by 6, both 2 and 3 must divide the number.

15.   a) 1, 4, 7

         b) 826,3<u>20</u>   826,3<u>40</u>   826,3<u>60</u>   826,3<u>00</u>

         c) 417,2<u>10</u>   417,2<u>40</u>   417,2<u>70</u>

         d) 7,4<u>11</u>

17.   No

19.   a) False

b) 5 divides 5 but 10 does not divide 5

c)

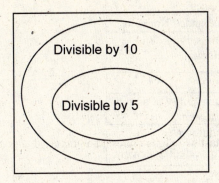

21.   False $6 \mid 24$ and $8 \mid 24$ but $48 \nmid 24$

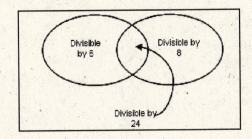

23.   1, 3, and 5

25. $(100 \times A) + (10 \times B) + (1 \times C)$ $10B$, and C ; $100A + 10B + C$

27.   It's a perfect cube, a perfect square, has a remainder of 1 when divided by 2. It is divisible by 3

29.

Statement	Reason
1. ABCD is a four digit number	Given
2. $ABCD = 1000A + 100B + 10C + D$	Expanded Notation
3. $ABCD = 999A + 99B + 9C$ $+ A + B + C + D$	Equivalent Statement
4. $9 \mid 999A, 9 \mid 99B,$ and $9 \mid 9C$	Divisibility of a Product
5. $9 \mid A + B + C + D$	Given
6. $9 \mid 999A + 99B + 9C$ $+ A + B + C + D$	Div. of a Sum Theorem
7. $9 \mid 1000A + 100B +$ $10C + D$	Addition
8. $9 \mid ABCD$	Equivalent Statement

31.

Statement	Reason
1. ABCD is a four digit number	Given
2. $ABCD = 1000A + 100B + 10C + D$	Expanded Notation
3. $4\mid1000$ then $4\mid1000A$ $4\mid100$ then $4\mid100B$	Divisibility of a Product
4. $4\mid10C + D$	Given
5. $4\mid1000A + 10B + (10C + D)$	Divisibility of a Sum Theorem
6. $4\mid1000A + 100B + 10C + D$	Addition
7. $4\mid ABCD$	Equivalent Statement

33.   a) $100\mid N$ if and only if the last two digits are 0.

b) (1) (3) (4)

35.   If a number is divisible by 3 and 5 then it is divisible by 15.

37.   5, 17, 29, 41,…

39.   a)   72 Yes, difference is divisible by 9
$$\begin{array}{r} 72 \\ -27 \\ \hline 45 \end{array}$$

b)   83 Yes, difference is divisible by 9
$$\begin{array}{r} 83 \\ -38 \\ \hline 45 \end{array}$$

c) Let $10x + y$ represent an arbitrary number. $10x + y - (10y + x) = 9x - 9y = 9(x - y)$ which must be

divisible by 9.

41.   a) The last digit must be 0 or 4

b) For 2, the last digit must be 0, 2, 4, or 6

For 7, the sum of the digits must be divisible by 7.

43.   a) 147147

        b) 147147

        c) Yes

        d) Yes

        e) Yes

        f) 216216

        g) 1001

        h) 382382

        i) Any number in the form ABCABC is divisible by 1001 so it must be divisible by 7, 11, and 13

4.3 Lesson Exercises

LE1   Opener

       A prime number has exactly two distinct factors one and itself.

LE2   Concept

       a) One

       b) Neither

LE3   Reasoning

a)

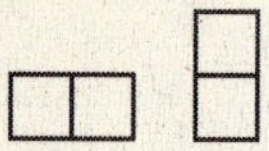

b)

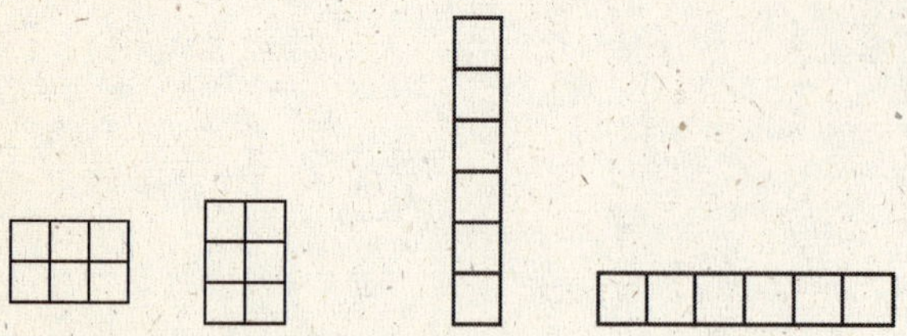

c)

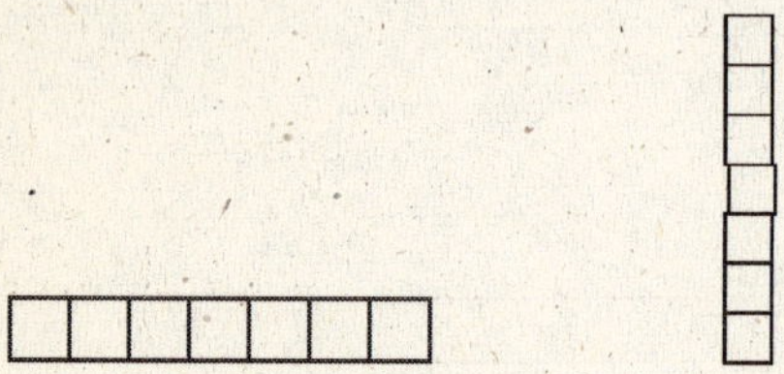

d)

e) A prime number can be represented by exactly two different rectangles with lengths that are counting

numbers.

f) Induction

LE4   Concept

2, 3, 5, 7, 11, 13, 17, 19

LE5   Opener

If 367 is not divisible by 2, 3, 4, 5, …365, 366 it is prime

LE6   Reasoning

a) Multiples of 2

b) Multiples of 3

LE7   Skill

a) 347; $\sqrt{347} = 18.6$, check 2, 3, 5, 7, 11, 13, 17

None of these numbers divide 347 so it is prime

b) 253; $\sqrt{253} = 15.9$ check 2, 3, 5, 7, 11, 13

$11 | 253$ so it is not prime

LE8   Opener

a) Yes

b) No

d) A composite number can be written as a unique product of primes

e) Induction

LE9   Skill

Factor tree method

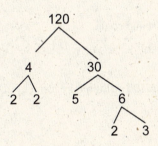

$120 = 2^3 \bullet 3 \bullet 5$

Prime divisor method

$2 | 120$
$2 | 60$
$2 | 30$ $120 = 2^3 \bullet 3 \bullet 5$
$3 | 15$
$\quad 5$

LE10   Communication

Each prime number at the end of the limbs in a factor tree must be included in the prime factorization.

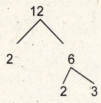

$$12 = 2^2 \cdot 3$$

LE11   Concept

a) 3 and 5, 5 and 7, 11 and 13, 17 and 19, 29 and 31, 41 and 43, 59 and 61, 71 and 73

b) They are all divisible by 6

c) Yes it seems like there is an infinite number of twin primes

LE12   Concept

a) $8 = 3 + 5$

b) $22 = 11 + 11$

c) $120 = 47 + 73$ or $7 + 113$ or $11 + 109$ or $13 + 107$ or $19 + 101$ and so on

LE13   Reasoning

This is inductive reasoning. Some numbers even greater than 100 million may turn out to be a counter example.

LE14   Summary

A counting number greater than 1 is prime if it has exactly two distinct factors. To determine if a number N is prime find out if any prime number less than or equal to $\sqrt{n}$ is a divisor of N. Every composite number has exactly one prime factorization.

4.3 Homework Exercises

1. a) prime

    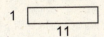

    b) composite

3. Ex $4 \times 5 = 20$ and $20 + 17 = 37$ all results will be prime

5. a) $2^{2^3} + 1 = 257 = 4th$ Fermat number

    b) $2^{2^5} + 1 = 2^{32} + 1 = 2^{16} + 2^{16} + 1 = 4,294,967,297$ and
    $$4,294,967,297 \div 641 = 6,700,417$$

7. a) It is divisible by 3, 5, 7, 11, 13, etc.

    b) It is divisible by 2

    c) It is divisible by 5

9. $\sqrt{431} \sim 20.7$ Try 2, 3, 5, 7, 11, 13, 17, 19

11. a) $\sqrt{17} \sim 8.4$ Try 2, 3, 5, 7, Prime

    b) $\sqrt{697} \sim 26.4$ Try 2, 3, 5, 7, 11, 13, 17, 19, 23 Composite $17 \bullet 41 = 697$

13. a) A whole number that has all 3's except for a last digit of 1 is a prime number

    b) Inductive

    c) Composite divisible by 17

15. Exactly one prime factorization

17.   a) Least Prime

$3\underline{|495}$
$3\underline{|165}$ $= 3^2 \bullet 5 \bullet 11$
$5\underline{|55}$

Factor Tree

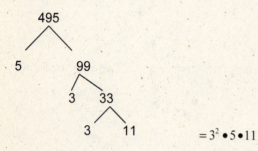

$= 3^2 \bullet 5 \bullet 11$

b) Least Prime

$2\underline{|320}$
$2\underline{|160}$ $= 2^6 \bullet 5$
$2\underline{|55}$
$2\underline{|40}$
$2\underline{|20}$
$2\underline{|10}$
5

19.   a) $6 \bullet 3 \bullet 2 = 36$ divisors

b) $148 = 2^2 \bullet 37$ It has $3 \bullet 2 = 6$ divisors

21.   101 and 103, 107 and 109, 137 and 139

23.   a) $12 = 5 + 7$

b) $30 = 13 + 17$ or $11 + 19$ or $23 + 7$

c) $108 = 103 + 5$

25.   2 divisors: all prime

3 divisors: 9, 25

4 divisors: 8, 10, 14, 15, 21, 22

5 divisors: 16

6 divisors: 12, 18, 20

8 divisors: 24

Perfect square numbers have an odd number of divisors. Prime numbers have two divisors. Squares of odd

numbers greater than 1 have 3 divisors

27.   a) $2|722$

b) $3|723$

c) $4|720$ and $4|4$ so $4|724$

d) $5|720$ and $5|5$ so $5|725$; $6|720$ and $6|6$ so $6|726$

29.   a) $x = 3$

b) No solution

c) $x = 7$

d) x is the greatest prime number less than the divisor, but it doesn't work all the time.

31.   a)

g) 2, 3, 5, 7, 11, 13, 17, 19, 23, 29, 31, 37, 41, 43, 47

4.4 Lesson Exercises

LE1   Opener

a) For the student

b) Common factors of 20 and 30: 1, 2, 5, 10

c) 10

d)

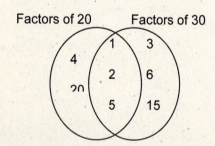

LE2   Concept

The counting number

LE3   Skill

$546 = 2 \bullet 3 \bullet 7 \bullet 13$ and $234 = 2 \bullet 9 \bullet 13$ then $GCF(546, 234) = 2 \bullet 13$
$$= 26$$

$F_{546} = \{1, 2, 3, 6, 7, 13, 14, 21, 26, 39, 42, 78, 91, 182, 273, 546\}$

$F_{234} = \{1, 2, 9, 13, 18, 26, 117, 234\}$

$GCF(546, 234) = 26$

LE4   Communication

The GCF is the largest number that is a factor of both 60 and 140. What number besides 5 are common

factors of 60 and 140? Is 2.5 a common factor? What is the greatest common factor?

LE5   Skill

b and c

LE6   Reasoning

a) Yes

b) Inductive

LE7   Skill

GCF (253, 322) = GCF (69, 253) = GCF (46, 69) = GCF (23, 46) = 23

$$
\begin{array}{cccc}
1 & 3 & 1 & 2 \\
253\overline{)322} & 69\overline{)253} & 46\overline{)69} & 23\overline{)46} \\
\underline{253} & \underline{207} & \underline{46} & \underline{46} \\
69 & 46 & 23 & 0
\end{array}
$$

LE8   Opener

a) For the student

b) 40, 80, 120

c) 40

LE9   Reasoning

a) Yes

b) Yes

c) MN

d) Inductive

e) MN is a whole number time M, so it is a multiple of M. MN is a whole number times N, so it is a

multiple of N.

LE10   Skill

LCM (120, 72)

Multiple List

72: 72, 150, 216, 288, 360 …

120: 120, 240, 360 …

LCM (120, 72) = 360

Prime Factorization

LCM (120, 72)

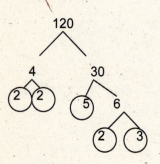
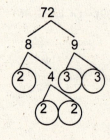

$120 = 2^3 \bullet 3 \bullet 5$
$72 = 2^3 \bullet 3^2$

$LCM(72,120) = 2^3 \bullet 3^2 \bullet 5$
$\qquad\qquad\quad = 360$

LE11   Skill

2,480

LE12   Reasoning

a) When GCF (M, N) = 1.

b) For the student

c) When M and N are relatively prime.

LE13   Summary

a) To find the GCF, one can use the factor list method, prime factorization method or Euclidean Algorithm.

b) To find the LCM, one can use either the multiple list method or the prime factorization method.

4.4 Homework Exercises

1.   a)

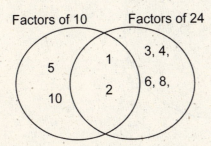

b) GCF (10, 24) = 2

3.   a)

42: 1, 2, 3, 6, 7, 14, 21, 42

120: 1, 2, 3, 4, 5, 6, 8, 10, 12, 15, 20, 24, 30, 60, 120

GCF = 6

b)

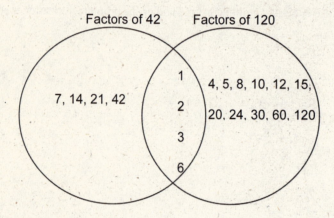

c) $42 = 2 \cdot 3 \cdot 7$ $GCF = 2 \cdot 3 = 6$

   $120 = 2^3 \cdot 3 \cdot 5$

5. $F_{96} = \{1, 2, 3, 4, 6, 8, 12, 16, 24, 32, 48, 96\}$ and $F_{120} = \{1, 2, 3, 4, 5, 6, 8, 10, 12, 15, 20, 24, 30, 40, 60, 120\}$

   $GCF(96, 120) = 24$

   $96 = 2^5 \cdot 3$ and $120 = 2^3 \cdot 3 \cdot 5$ so $GCF(96, 120) = 2^3 \cdot 3 = 24$

7.   a) GCF (4, 5) = 1 GCF (4, 6) = 2 GCF (4, 7) = 1 GCF (4, 8) = 4 GCF (4, 9) = 1

   GCF (4, 10) = 2 GCF (4, 11) = 1 GCF (4, 12) = 4

   b) The sequence of GCF's will have the repeating pattern 1, 2, 1, 4, 1, 2, …

   c) Induction

9.   The GCF must also include 3 and 5. It is $3 \cdot 5 \cdot 11 = 165$

11. $GCF(a, b) = 2^2 \cdot 3 \cdot 5^6$

13.   a, c

15. a) GCF (627, 665): $627\overline{)665}^{\,1}$ $38\overline{)627}^{\,16}$ $19\overline{)38}^{\,2}$ GCF (627, 665) = 19

$$\underline{672}$$
$$38$$

$$\underline{38}$$
$$247$$
$$\underline{228}$$
$$19$$

$$\underline{38}$$
$$0$$

b) GCF (851, 2035): $851\overline{)2035}^{\,2}$ $333\overline{)851}^{\,2}$ $185\overline{)333}^{\,1}$ $148\overline{)185}^{\,1}$ $37\overline{)148}^{\,4}$

$$\underline{1702}$$
$$333$$

$$\underline{666}$$
$$185$$

$$\underline{185}$$
$$148$$

$$\underline{148}$$
$$37$$

$$\underline{148}$$
$$0$$

GCF (851, 2035) = 37

c) GCF (551, 609): $551\overline{)609}^{\,1}$ $58\overline{)551}^{\,9}$ $29\overline{)58}^{\,2}$ GCF (551, 609) = 29

$$\underline{551}$$
$$58$$

$$\underline{522}$$
$$29$$

$$\underline{58}$$
$$0$$

17. a) 60, 120, 180

b) An infinite number

c) 60

19. a)

20: 20, 40, 60, 80, 100, 120, 140, 160

32: 32, 64, 96, 128, 160

LCM = 160

b) $20 = 2^2 \bullet 5$ $LCM = 2^5 \bullet 5 = 160$
    $32 = 2^5$

21.   a)

Multiples of 18	Multiples of 40
18	40
36	80
54	120
72	160
90	200
108	240
126	280
144	320
162	360
180	400
198	440
216	480
234	520
252	560
270	600
288	640
306	680
324	720
342	760
360	800

$LCM(18, 40) = 360$

b) $LCM(18, 40) = 2^3 \bullet 3^2 \bullet 5 = 360$

23.   a) LCM (8, 10, 24) = 120

b) 240, 360, 480, …

25.   a)

b) 6 in by 6 in GCF (30, 48) = 6

27.   a) Yes

b) No

29.   a) $LCM(a,b) = 2^3 \cdot 3^4 \cdot 5^7 \cdot 7 \cdot 11$

b) $GCF(a,b) = 2 \cdot 3^2 \cdot 5 \cdot 11$

31. $b = 2 \cdot 3^2 \cdot 5^4 \cdot 7$

33.   b

35. $3y$

37.   False   A = 7 B = 9

39.   b and c

41.   a)

a	b	$GCF(a,b)$	$LCM(a,b)$	$GCF(a,b) \cdot LCM(a,b)$
3	4	1	12	$1 \cdot 12 = 12$
6	8	2	24	$2 \cdot 24 = 48$
5	15	5	15	$5 \cdot 15 = 75$

b) $a \cdot b = GCF(a,b) \cdot LCM(a,b)$

43.   For student

Review Exercises Chapter 4

1. Write the hypotenuse as the first step and the conclusion as the last step. Then convert the hypothesis and conclusion into equations and write these as the second and next to last steps, respectively. Finally, use the properties of equations to work from the second step to the next to the last step.

3.   6   0, 3, 6, 9 followed by 0.

            3

    $32,005,0\underline{20}$

           9

5. $\sqrt{577} \sim 24$ Try 2, 3, 5, 7, 11, 13, 17, 19, 23

7. $1911 = 3 \bullet 7^2 \bullet 13$

9. True $A = 2$, $B = 4$ is an example

11. False   10 is divisible by 10 but 10 is not divisible by 20

13. False   2 and 4 is an example

15. Two possible answers are $A = 3$ or $A = 21$.

17. $24x = 168(2)$

      $x = 14$

19. Using prime factorization, 5 and 7 are prime, $6 = 2 \bullet 3$ and $8 = 2^3$, the $LCM(5,6,7,8) = 2^3 \bullet 3 \bullet 5 \bullet 7 = 840$

Chapter Five

Lesson Exercises 5.1

LE1   Opener

a) stock market, debit in your checkbook

b) $x + 3 = -1$ solution is a negative number

LE2   Communication

$|-13|$ represents the distance -13 is away from zero on the number line. Both 13 and -13 are thirteen units

away from zero.

LE3   Opener

a) 5 degrees below 0

b) 4 under par

c) a debt of $400

d) a decrease of $3 / share

LE4   Skill

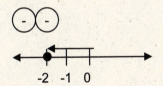

LE5   Concept

a) -132

b) $-7, -2, 1, 3$

LE6   Communication

a)

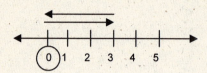

Go to 3. To add -3, move 3 units left, you end up at 0. So, $3 + (-3) = 0$

b)

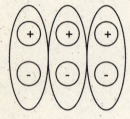

Show 3 as 3 positive counters and -3 as 3 negative counters. Form 3 zero pairs. There are no other counters

left, so $3 + (-3) = 0$.

c) 0

LE7   Communication

a) Go to 4. Then move 1 to the left

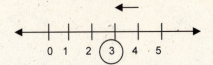

You end up at 3. So $4 + (-1) = 3$

b) Show 4 as 4 positive counters and -1 as 1 negative counter.

a)

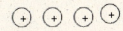

Forms one zero pair. This leaves 3 positive counters. So $4 + (-1) = 3$

c)

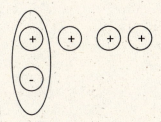

The answer comes out positive because 4 has a larger absolute value than -1.

LE8   Communication

To add two negative integers, add their absolute values and make the result negative.

LE9   Communication

Difference; larger absolute value

LE10   Concept

The difference of 4 and -1 is three. Since 4 has the larger absolute value the positive sign is kept. $4 + -1 = 3$

LE11   Connection

a) $\$282 + (-\$405) = -\$123$

b) measures/groups

LE12   Connection

I owe Sue $10 and just received a bill today for $14. How much is the total that I owe? $-10+(-14)=-24$

LE13   Opener

a)

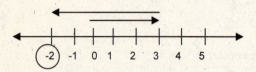

b) I have $3 dollars, but owe Judy $5.

c) $3-5=3+-3+-2=0+-2=-2$

d)

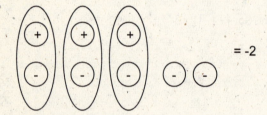

$= -2$

LE14   Reasoning

a) Right

b) Left

c) Left

d) Right

LE15   Communication

a) Go to -2, move 1 to the left. You end up at -3. So $-2-1=-3$

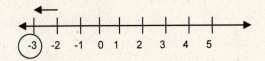

b) Go to 2. Adding -3 would move to the left so subtracting -3 moves 3 to the right. You end up at 5. So

$2-(-3)=5$

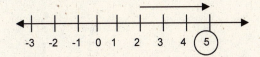

c) Go to -3. Adding -2 is a move to the left, so subtracting -2 is a move of 2 to the right. You end up at -1.

So $-3-(-2)=-1$

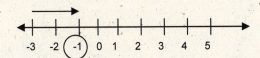

LE16   Concept

a) No difference

b) 5; -5

c) -15; -15

d) No difference

e) -5; 5

f) -5; -5

LE17   Reasoning

a) In subtracting a positive integer a, you move a units to the left, and in adding $-a$, you move a units to

the left.

b) In subtracting a negative integer $-a$, you move a units to the right, and in adding a, you move a units to

the right.

c) Subtracting a positive numbers is the same as adding its opposite. Subtracting a negative number is the

same as adding its positive.

LE18   Skill

a) 926

b) -288

LE19   Connection

a) $14 - 38 = -24$

b) $-24^0\,\text{C}$

c) measures

d) take away

LE20   Communication

It is 2 degrees below zero. During the next hour, the temperature drops 6 more degrees. What is the

temperature now?

$-2 - 6 = -8$

LE21   Summary

Integer addition and subtraction can be effectively modeled for children using a number line and signed

counters.

5.1 Homework Exercises

-1.   a, b, d

0.   -1

1.   a) $N \cup W = I$

b) { }

3.   -5 and 5 are the same distance (5) from 0

5.   a) loss of yardage

b) more expenditures than receipts

c) below sea level

7.   -6 means 6 less than the normal high

7 means 7 more than the normal high

9.

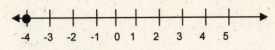

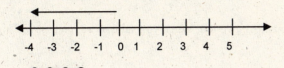

Or

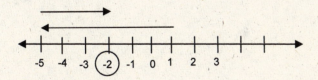

11.   a) 20 seconds before liftoff; -20

b) a gain of $3; 3

c) 3 floors lower; - 3

13.   a) What number is 3 units to the right of -100? -97

b) -8,-2,1,5

15. $-4+-2=-6$

17.   a)

b) combine measures

c)

= -2

19.   a) $\$86+(-\$30)+(-\$20)=\$36$

b) Combine groups/measures

21.   Show the student that -5 is to the right of -10 on the number line, making it the larger number.

23.   a) -29   b) 365   c) -108

25.   b

27.   a) move right 4 from zero then move 6 left from 4 which equals -2.

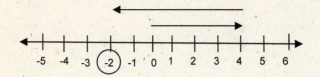

b) Move right 5 from zero, when you add -2 you move two left. When you subtract -2, you move 2 right

ending at 7.

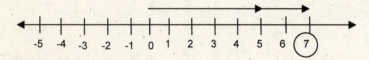

29.   a) 3, left

b) 3, left

c) Adding $-N$ is the same as subtracting N, where N is an integer.

31.   a) -73   b) 38

33.   The temperature at 11 p.m. was -6 and during the night the temperature dropped 4 degrees, what was the

new temperature? $-6 + -4 = -10$

35.   a) $-2 + 3 - (-5) = 6$

b) $2 - 3 + -5 = -6$

c) $-2 - (-3) + (-5) = -4$

37.   a) -14   b) 8   c) 10

39.   a) $6194 - (-86) = 6280$m

b) $5642 - -28 = 5670$

c) Compare measures

41.

Temperature at 8 a.m.	Temperature at 6 p.m.	Change
7	-3	-10
-6	-2	-4
5	-3	8
-2	-4	-2
14	3	11

43.	a) -1, -2

b) deductive

c) $3 - 2 = 1$
$3 - 1 = 2$
$3 - 0 = 3$
$3 - (-1) = 4$
$3 - (-2) = 5$

d) $3 - 1 = 2$
$2 - 1 = 1$
$1 - 1 = 0$
$0 - 1 = -1$
$-1 - 1 = -2$
$-2 - 1 = -3$

45.	a) $6 = -2 + n; n = 8$

b) $-3 = 2 + n; n = -5$

47.	a) All odd integers

b) All positive and negative multiples of 3

49.	a) True $x = 3, y = 2$

b) False if $x < y$

51.

-2	-4	6
8	0	-8
-6	4	2

53.	For student

55.	For student.

Lesson Exercises 5.2

LE1   Opener

Take two negative counters three times.

How many negative counters do you have? Ans: -6

LE2   Concept

a) $3 \bullet (-2) = -2 + (-2) + (-2) = -6$

b) $3 \bullet (-2)$ means 3 sets of -2.

This makes -6 so $3 \bullet (-2) = -6$

c)

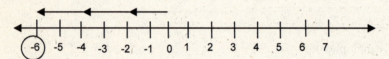

d) Negative integer

LE3   Reasoning

a) $(-2) \bullet 3 = 3 \bullet (-2) = -6$

b) Negative integer

LE4   Reasoning

a) 2; 4; 6

b) Positive

LE5   Concept

a) $-6 \div 3 = -2$ as $-2 \times 3 = -6$

b) $6 \div -3 = -2$ as $-3 \times 2 = -6$

c) $-6 \div -3 = 2$ as $-3 \times 2 = -6$

d) Positive

e) Negative

The product of two integers with the same sign is positive

The product of two integers with different signs is negative

The quotient of two integers with the same sign is positive

The quotient of two integers with different signs is negative

In multiplication if the two factors have like signs the product is positive, if the two factors have different

signs the product is negative.

Homework Exercise 5.2

1.   a) $-3 + -3 + -3 + -3 = -12$

       b)

       c)

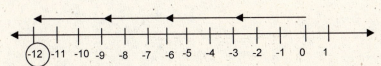

3.   a) A negative times a positive equals a negative

       b) Inductive

5. $-5 \bullet -1 = 5$

       $-5 \bullet -2 = 10$

7.   a) $+12$ lbs.

       b) $-3 \times 4 = -12$

       c) Equal measures

9.   a) $-6 \times \underline{\quad} = -54; 9$

       b) $-4 \times \underline{\quad} = 32; -8$

11.   a) $\$-40,000$

       b) $-480000 \div 12 = -40000$

       c) Partition a group/measures

13.   a) -54   b) 240   c) -3   d) 1

15.　　a) -7　　b) -37

17.　　a) $t = 0$

　　b) $y = 2$ or -2

　　c) -5

19.　　a) Positive　　b) Positive　　c) Negative

21.　　Gain of 2

23.　　First Row:　　$1 \times (-4) \times 2$

　　2nd Row:　　$-2 - (-4) - (-2)$

　　3rd Row:　　$2 \div (-2) \div (-1)$

25.　　For all x and y

27.　　$(-2 \div -2) + (-2 \div -2) = 2$
　　$(-2 \times -2) - (-2 \div -2) = 3$
　　$(-2 \times -2) + -2 - (-2) = 4$
　　$(-2 \times -2) + (-2 \div -2) = 5$
　　$-2 \times [-2 - (-2 \div -2)] = 6$
　　$(-2 \times -2) + (-2 \div -2) = 8$
　　$(-2 + -2) \times -2 + (-2 \div -2) = 9$

Lesson Exercises 5.3

LE1　　Opener

　　Commutative, associative, distributive, additive identity, multiplicative identity, closure with respect to

　　addition and multiplication

LE2　　Connection

　　a) Add $-5 + 5$, then add 3

　　b) Associative property of addition

LE3　　Skill

　　$l(-5 + 2) = -3l$

LE4   Skill

a) Commutative

b) If a rule does not apply to all whole numbers, then it cannot apply to all integers, since every whole number is also an integer. Whatever was a counter example for whole numbers is also a counter example for the set of integers.

LE5   Skill

a) -3; -3; -3

b) 8; 8; 8

c) Yes

LE6   Skill

a) -7

b) 2

c) 0

LE7   Opener

Yes

LE8   Reasoning

a) 4; 2

b) The child computes the larger number minus the smaller.

LE9   Reasoning

a) -20; -18

b) The child thinks a negative number times a negative number is a negative number.

LE10   Reasoning

a) -9; -10

b) The child adds a negative number and a positive number by adding their absolute values and placing a negative sign in front of the result.

LE11   Skill

-2

LE12   Reasoning

a) Plan, check a couple of values for $a = 5, b = 3$

b) Carry out the plan with values from (a) $a - b = 5 - 3 = 2$
$$b - a = 3 - 5 = -2$$

d) Generalization $a - b$ and $b - a$ are additive inverses of each other as $a - b + b - a = 0$

e) Inductive

LE13   Reasoning   When $m = n$

LE14   Summary

Integer addition, subtraction, and multiplication are closed. All the whole number properties also apply to integers as every integer is also a whole number. The sum of an integer x and its inverse $(-x) = 0$.

$x + -x = 0$. Division is not closed for integers, for example $6 \div 12 = \dfrac{1}{2}$, this quotient is not an integer.

Homework Exercise 5.3

1.   a) Addition and multiplication

b) Addition and multiplication

3. $(-5 \times -8) \times 7$

5.   a) -3

b) 11

c) Commutative and associative properties of addition

7.   Commutative, multiplication

9.   a) Associative and distributive properties

b) The distributive property distribute multiplication over addition or subtraction. This problem only involves multiplication.

11. $(10 \div 2) \div 5 \overset{?}{\neq} 10 \div (2 \div 5)$
$$5 \div 5 \overset{?}{\neq} 10 \div \dfrac{2}{5}$$
$$1 \neq 25$$

13. a) Any counter example showing that whole number division is associative would also show integer division is not associative.

b) Deductive

15. 1, identity element, multiplication

17. 0

19. Closure property of subtraction

21. a) 5, 7

b) The unreadable

23. Mixing up the addition and multiplication rules

25. a) $a \div b = 1/(b \div a)$

b) When $a = b$

27. a) $0 = 2 \bullet 4 + (-2 \bullet 4)$

b) $0 = 2 \bullet 4 + (-2 \bullet 4) = 8 + (-2 \bullet 4)$ Therefore $(-2 \bullet 4) = -8$ because each integer has a unique additive inverse

29. $(2m)(2n) = 4mn = 2(2mn)$ Which is even

Chapter 5 Review Exercises

1. The set of whole numbers is $\{0, 1, 2, 3, ...\}$. The set of whole numbers is a subset of the set of integers. The set of integers is the union of the set of whole numbers and their opposites $\{..., -3, -2, -1, 0, 1, 2, 3, ...\}$

3. a) Show -4 as 4 negative counters. Now take away -2.

2 black counters are left so $-4 - (-2) = -2$

b)

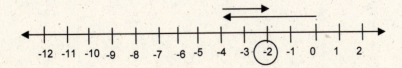

5.   a) 500 soldiers better

        b) $-300 - (-800) = 500$

        c) Compare groups

7.   a)

 = -12

        b)

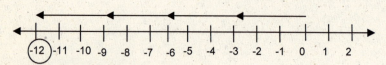

9. $-18 \div -3; -3 \times \underline{\quad} = -18$

11.   a) -37   b) -2   c) 157   d) -9

13. $n(-5 + 8)$

Chapter Six

Lesson Exercises 6.1

LE1   Reasoning

It draws attention to the sign.

LE2   Opener

$\frac{2}{3}$ can be part of a whole or group. It can be a point on a number line or represent a division problem $2 \div 3$

LE3   Concept

$\frac{3}{4}$ represents

Part of a whole (shade 3 of 4 equal parts)

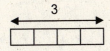

Part of a set (shade 3 of 4 equal groups)

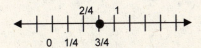

$3 \div 4$

Division (divide 3 into 4 equal parts)

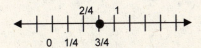

Point on a number line (go $\frac{3}{4}$ of the way from 0 to 1)

LE4   Connection

a) (3)

b) (2)

c) (1)

d) (4)

LE5   Connection

Divide each brownie into 2 equal parts or give each person 2 brownies and divide the remaining one into

two parts.

LE6   Reasoning

a)

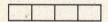

b)

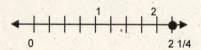

LE7   Communication

A fraction represents a relationship between equal parts and a whole. It does not compare the size of parts

from two different sized wholes.

LE8   Concept

a) 9

b) 2 wholes are 2×4 quarters plus 1 more quarter makes 9 quarters or $\dfrac{9}{4}$

c)

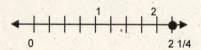

d) $\dfrac{9}{4} = 9 \div 4 = 2\dfrac{1}{4}$

LE9   Concept

(a) (b) (d) (e) (f)

LE10   Concept

a)

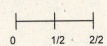

0   1/2   2/2

b)

0   2/4   4/4

c) They are all the same distance away from zero.

d) You obtain an equivalent fraction

LE11   Concept

$$\frac{6}{8} = \frac{6 \div 2}{8 \div 2} = \frac{3}{4}$$

LE12   Reasoning

No, a counterexample is $\dfrac{3}{4} \neq \dfrac{3+2}{4+2}$

$$= \frac{5}{6}$$

LE13   Reasoning

a) 4

b) 14

c) Greatest common factor $14 = \text{GCF}(28, 42)$

d) To get a/b in simplest form divide both a and b by the $\text{GCF}(a, b)$

LE14   Skill

$$\text{GCF} = 4; \quad \frac{148}{260} = \frac{148 \div 4}{260 \div 4} = \frac{37}{65}$$

LE15   Skill

$\dfrac{2}{3}$ ▨▨▨▨░░

$\dfrac{5}{6}$ ▨▨▨▨▨░

$\dfrac{2}{3} < \dfrac{5}{6}$

LE16   Skill

a) $\dfrac{18}{30} = \dfrac{3}{5}$ and $\dfrac{24}{40} = \dfrac{3}{5}$ so $\dfrac{18}{30} = \dfrac{24}{40}$

b) $\dfrac{2}{3} = \dfrac{10}{15}$ and $\dfrac{3}{5} = \dfrac{9}{15}$ so $\dfrac{2}{3} > \dfrac{3}{5}$

LE17   Reasoning

a) $\dfrac{19}{30}$

b) $\dfrac{37}{60}$ and $\dfrac{39}{60}$ $\left(\text{or } \dfrac{13}{20} \right)$

c) An infinite number

LE18   Summary

There are many ways to explain a fraction to a student. The following are good representations:

1) Fraction of a whole

2) Fraction of a set

3) Number line

4) Division

Homework Exercises 6.1

1.   a)

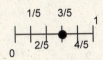

       b)

3.   a) 1 brownie, $2 \div 2 = 1$

       b) $\dfrac{2}{3}$ brownie $2 \div 3 = \dfrac{2}{3}$

       c) $\dfrac{1}{2}$ brownie $2 \div 4 = \dfrac{1}{2}$

5.   The figure is not divided into 5 <u>equal</u> parts.

7.   Part of a whole

$\frac{1}{3}$

Part of a set

$\frac{1}{3}$

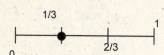

Division

$$3\overline{)1.0}^{\;.3}$$

$\frac{1}{3}$

Number line

$\frac{1}{3}$

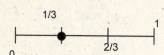

9.   a)

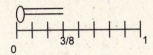

0   3/8   1

b)

c)

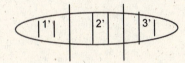

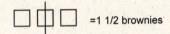

d)

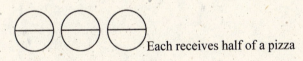

11.

☐ ⊟ ☐   =1 1/2 brownies

or

⊟ ⊟ ⊟   3 ½ = 1 1/2 brownies

13.   a)

◯ ◯ ◯ Each receives half of a pizza

b)

Each person receives $\frac{1}{8}$ of a pizza. Since there are 3 pizzas each receives 3 "eighths."

15.   The remainder 25 what part of the division 50? It is $\frac{25}{50}$ or $\frac{1}{2}$.

17. $\frac{0}{5} = 0$ Because $5 \bullet 0 = 0$

19. a)

b)

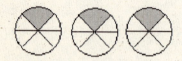

c) $\frac{1}{3}$

21. a) All of them

b) $\frac{-5}{12} = -\frac{5}{12} \bullet \frac{(-1)}{(-1)} = \frac{5}{-12} = -\frac{5}{12}$

23. The whole is represented as:

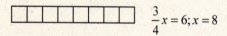 $\frac{3}{4}x = 6; x = 8$

25.

$\frac{3}{5}x = 6, x = 10$ cats

27. Three $\frac{3}{8}$ could be represented as:

So the whole could represented as:

4 rectangles

29.

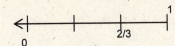

31.   This is correct if the X in the illustration represents $\dfrac{1}{6}$, therefore three X's is one-half of the whole. If on the

other hand three X's represent $\dfrac{1}{3}$ then the student is incorrect, as the students needs to understand that

$\dfrac{5}{6} < 1$.

33.

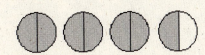

They are both equal to 1 pizza, no matter how many slices they are cut into.

35.   a) Part of a whole meaning $3\dfrac{1}{2} = \dfrac{7}{2}$

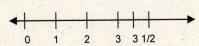

$1 + 1 + 1 + \dfrac{1}{2} = 3\dfrac{1}{2} = \dfrac{7}{2}$

b) $3\dfrac{1}{2}$ when written in terms of a common denominator is $\dfrac{6}{2} + \dfrac{1}{2} = \dfrac{7}{2}$. In 3 there are six (halves) plus one

more half $= 7$ (halves).

c)

$\begin{array}{ccccc} & | & | & | & || \\ 0 & 1 & 2 & 3 & 3\,1/2 \end{array}$

d) $4\overline{\smash{)}9} = 2\dfrac{1}{4}$
$\underline{-8}$
$\;1$
with quotient 2

37.   a) $\dfrac{3}{1}$ b) $\dfrac{-3}{1}$ c) $\dfrac{9}{2}$ d) $\dfrac{-56}{10}$ e) $\dfrac{25}{100}$

39.   a) Yes   b) W

41.   a) $\dfrac{3}{5}$

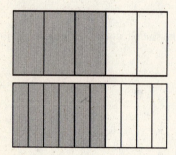

$\dfrac{6}{10}$

Represent the same amount

     b)

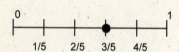

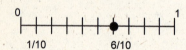

     c) $\dfrac{3}{5} \times \dfrac{2}{2} = \dfrac{6}{10}$

43. $\dfrac{1200}{1500} = \dfrac{4}{5}$ $\dfrac{4}{5}$ of $2000 = 1600$

45.   The child is "correct" that each fraction is one part less

than a whole, but the child needs to understand that

the "one part" is not the same. One part of 4 pieces is not the

same as one part of 8 pieces

47.   a) $\dfrac{21}{58}$

   b) $\dfrac{1}{yz}$

49.   a)

   b) $\dfrac{5}{8} = \dfrac{15}{24}$

   $\dfrac{1}{3} = \dfrac{8}{24}$

   c) $\dfrac{4}{8} = \dfrac{1}{2}$ so $\dfrac{5}{8}$ is greater than one-half

   $\dfrac{1}{3} = \dfrac{2}{6}$ so $\dfrac{1}{3}$ is less than one-half
   $\dfrac{1}{2} = \dfrac{3}{6}$

51. $\dfrac{4}{9} = \dfrac{20}{45}$ $\dfrac{7}{15} = \dfrac{21}{45}$ $\dfrac{4}{9} < \dfrac{7}{15}$

53. $\dfrac{14}{23} = \dfrac{420}{690}$ $\dfrac{17}{30} = \dfrac{391}{690}$ $\dfrac{14}{23} > \dfrac{17}{30}$

55. $-\dfrac{1}{2}$ is greater

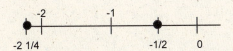

   $-\dfrac{1}{2}$ is to the right of $-2\dfrac{1}{4}$

57.

59. $\dfrac{3}{4} = \dfrac{6}{8} = \dfrac{24}{32}$ $\dfrac{7}{8} = \dfrac{28}{32}$ Answer: $\dfrac{25}{32}, \dfrac{26}{32}, \dfrac{27}{32}$

61. No

63. a) 6   b) $\dfrac{1}{6}$ c) $\dfrac{1}{2}$ d) $\dfrac{2}{3}$

65. No, they must express $\dfrac{5}{8}$ and $\dfrac{5}{9}$ in terms of a common denominator. Here $\dfrac{5}{8} = \dfrac{45}{72}$, $\dfrac{5}{9} = \dfrac{40}{72}$ between

$\dfrac{5}{8}$ and $\dfrac{5}{9}$ is $\dfrac{41}{72}, \dfrac{42}{72}, \dfrac{43}{72}, \dfrac{44}{72}$

67. For Student

69. For student

Lesson Exercises 6.2

LE1   Skill

a) $\dfrac{a+b}{c}$

b) $\dfrac{d-e}{f}$

LE2   Opener

Use fraction bar pictures to show $\dfrac{1}{2} + \dfrac{1}{3} \neq \dfrac{2}{5}$

LE3   Connection

a)

b) $\frac{1}{2}+\frac{1}{4}$ would equal the shaded region shown but we cannot name the sum unless we use a common

denominator (fourths)

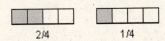

c)

d) $\frac{1}{2}+\frac{1}{4}=\frac{2}{4}+\frac{1}{4}=\frac{3}{4}$

3/4

LE4   Skill

a) $12x$

b) $\dfrac{9+10x}{12x}$, $x \neq 0$

LE5   Connection

a)

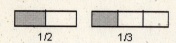

1/2   1/3

b) How much is left when we take $\dfrac{1}{3}$ away from $\dfrac{1}{2}$? We can't tell, a common denominator is needed.

c)

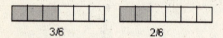

3/6   2/6

d) $\dfrac{1}{2} - \dfrac{1}{3} = \dfrac{3}{6} - \dfrac{2}{6} = \dfrac{1}{6}$

LE6   Skill

$$\frac{3}{20} - \frac{1}{12} = \frac{9}{60} - \frac{5}{60} = \frac{4}{60} = \frac{1}{15}$$

LE7   Connection

Subtraction, compare measures

LE8   Skill

$$2\frac{1}{2} + 6\frac{5}{6} = 2\frac{3}{6} + 6\frac{5}{6}$$
$$= 8\frac{8}{6}$$
$$= 8\frac{2}{6}$$
$$= 8\frac{1}{3}$$

LE9   Skill

a) $8\frac{2}{3}-3\frac{5}{6}=8\frac{4}{6}-3\frac{5}{6}=7\frac{10}{6}-3\frac{5}{6}=4\frac{5}{6}$

b) $8\frac{2}{3}-3\frac{5}{6}=\frac{26}{3}-\frac{23}{6}=\frac{52}{6}-\frac{23}{6}=\frac{29}{6}=4\frac{5}{6}$

c) The regrouping method takes fewer steps and avoids very large numerators. The improper fraction

method is easier to understand, since it does not require regrouping and is more like the methods for

multiplying and dividing mixed numbers.

LE10   Skill

For the student if you have a fraction calculator.

LE11   Skill

For the student if you have a fraction calculator.

LE12   Summary

Before one can add or subtract fractions with unlike denominators, a common denominator needs to be

found. Then each fraction must be written as an equivalent fraction in terms of the common denominator.

Homework Exercises 6.2

1.   a) $\frac{4}{y}$, $y \neq$ b) $\frac{-x}{n}$, $n \neq 0$

3.   Wrong

5.	a)

1/3   1/4

b) $\dfrac{1}{3}+\dfrac{1}{4}$ would equal the shaded region shown, but we cannot name it unless we use a common

denominator (twelfths)

c)

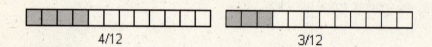

4/12   3/12

d) $\dfrac{1}{3}+\dfrac{1}{4}=\dfrac{4}{12}+\dfrac{3}{12}=\dfrac{7}{12}$

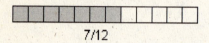

7/12

7.	$\dfrac{1}{3}$ is one third of the set ● ○ ○ and $\dfrac{1}{3}+\dfrac{1}{3}$ is represented by

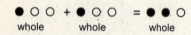

	● ○ ○ + ● ○ ○ = ● ● ○
	whole whole whole

	The whole is 3 circles not 6 circles

9.	a) $\dfrac{5}{12}+\dfrac{3}{8}=\dfrac{10}{24}+\dfrac{9}{24}=\dfrac{19}{24}$

	b) $\dfrac{4}{9}+\dfrac{2}{3}=\dfrac{12}{27}+\dfrac{18}{27}=\dfrac{30}{27}$

	c) $\dfrac{1}{a}+\dfrac{2}{b}=\dfrac{b}{ab}+\dfrac{2a}{ab}=\dfrac{b+2a}{ab},\ a\neq 0, b\neq 0$

11.	You could present b first as the denominators are simpler when thought of in terms of dollars. Then present

	c and d as these have common denominators then d and e as the denominators are not common.

13.   a)

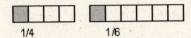

1/4   1/6

b) How much is left when we take $\frac{1}{6}$ away from $\frac{1}{4}$? We cannot tell. A common denominator is needed.

c)

3/12   2/12

d) $\frac{1}{4} - \frac{1}{6} = \frac{3}{12} - \frac{2}{12} = \frac{1}{12}$

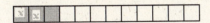

15.   a) $\frac{7}{20} = \frac{49}{140}$ $\frac{3}{28} = \frac{15}{140}$ 140 is the least common denominator

b) 280, 420

c) $\frac{7}{20} - \frac{3}{28} = \frac{49}{140} - \frac{15}{140} = \frac{34}{140} = \frac{17}{70}$

17.   a) $4\frac{1}{2} - 1\frac{3}{4} = 4\frac{2}{4} - 1\frac{3}{4} = 2\frac{3}{4}$ hours

b) Subtraction, take – away measures

19.   a) $5\frac{3}{4} + 2\frac{5}{8} = 5\frac{6}{8} + 2\frac{5}{8} = 7 + \frac{11}{8} = 7 + \frac{8}{8} + \frac{11}{8} = 8 + \frac{3}{8} = 8\frac{3}{8}$

b) $5\frac{3}{4} + 2\frac{5}{8} = \frac{23}{4} + \frac{21}{8} = \frac{46}{8} + \frac{21}{8} = \frac{67}{8} = 8\frac{3}{8}$

21.   a) $10\frac{1}{6} - 5\frac{2}{3} = 10\frac{1}{6} - 5\frac{4}{6} = 9\frac{7}{6} - 5\frac{4}{6} = 4\frac{3}{6} = 4\frac{1}{2}$

b) $\frac{61}{6} - \frac{17}{3} = \frac{61}{6} - \frac{34}{6} = \frac{27}{6} = 4\frac{3}{6} = 4\frac{1}{2}$

23.   You have a $5\frac{1}{4}$ hour job to do. You have been working for $2\frac{1}{2}$ hours. How much more time is needed?

$2\frac{3}{4}$ hours

25. You could present the problems a, c, b. Choice a is first due to common denominators, choice c is second since 5 and 10 are straightforward to finding a common denominator.

27. a) $\dfrac{1}{8}+\dfrac{4}{8}=\dfrac{5}{8}$ b) $3\dfrac{9}{10}-2\dfrac{6}{10}=1\dfrac{3}{10}$

29. a) $4\dfrac{1}{4}+3\dfrac{3}{4}=,4+3=7,\dfrac{1}{4}+\dfrac{3}{4}=1,7+1=8$

   b) $8-3\dfrac{1}{3}=7\dfrac{3}{3}-3\dfrac{1}{3}=4\dfrac{2}{3}$

31. a) $8\dfrac{1}{3}-3\dfrac{2}{3}=8\dfrac{2}{3}-4=4\dfrac{2}{3}$

   b) $5\dfrac{1}{4}-2\dfrac{3}{4}=5\dfrac{2}{4}-3=2\dfrac{2}{4}=2\dfrac{1}{2}$

33. a) $\dfrac{5}{12}$ b) $\dfrac{3}{20}$

35. a) $\dfrac{1}{4}=\dfrac{1}{5}+\dfrac{1}{20}$; yes

   b) $\dfrac{1}{N}=\dfrac{1}{N+1}+\dfrac{1}{N(N+1)}$

   c) Induction

   d) $\dfrac{1}{N}\overset{?}{=}\dfrac{1}{n+1}+\dfrac{1}{n(n+1)}$

   $\dfrac{1}{N}\overset{?}{=}\dfrac{n+1}{n(n+1)}$

   $\dfrac{1}{N}=\dfrac{1}{n}$

   e) Deduction

37. a) $\dfrac{1}{2}+\dfrac{1}{4}$

   b) $\dfrac{1}{26}+\dfrac{1}{13}$

   c) $\dfrac{1}{8}+\dfrac{1}{2}$

   d) $\dfrac{1}{9}+\dfrac{1}{2}+\dfrac{1}{6}$

Lesson Exercises 6.3

LE1   Concept

$$\underbrace{\frac{1}{8}+\frac{1}{8}+\frac{1}{8}+\frac{1}{8}+\frac{1}{8}}_{5\,\text{times}}=\frac{5}{8}$$

LE2   Communication

$\dfrac{1}{2}\times\dfrac{3}{5}$ is $\dfrac{1}{2}$ of $\dfrac{3}{5}$ show $\dfrac{3}{5}$

3/5

Now darken $\dfrac{1}{2}$ of $\dfrac{3}{5}$

3/5

$\dfrac{3}{10}$ of the figure is darkened so $\dfrac{1}{2}\times\dfrac{3}{5}=\dfrac{3}{10}$

LE3   Reasoning

a) $\dfrac{3}{8};\dfrac{3}{10}$ b) $\dfrac{ac}{bd}$ c) Induction

LE4   Opener

$\dfrac{3}{7}\times\dfrac{10}{21}=\dfrac{3\times10}{7\times21}$ is how we apply the definition for multiplication of fractions, but recall that the numbers 10

and 21 are not written in prime factored form, that is $10=2\bullet 5$ and $21=3\bullet 7$ so rewriting 10 and 21 in the

original problem yields, $\dfrac{3}{7}\times\dfrac{10}{21}=\dfrac{3\times10}{7\times21}$ Now we cross out the 3's since the quotient of 3 and itself is

$$=\frac{3\times2\times5}{7\times3\times7}$$

$$=\frac{3\times2\times5}{3\times7\times7}$$

$$=\frac{\cancel{3}\times2\times5}{\cancel{3}\times7\times7}$$

one.

LE5   Concept

$$\frac{4}{9} \times \frac{5}{8} = \frac{4 \times 1 \times 5}{9 \times 4 \times 2}$$

$$= \frac{4 \times 1 \times 5}{4 \times 2 \times 9}$$

$$= \frac{\cancel{4} \times 1 \times 5}{\cancel{4} \times 2 \times 9}$$

We can change the 4 and 8 to 1 and 2 respectively after we have rewritten $4 = 4 \times 1$ and $8 = 4 \times 2$. The

commutative law allows us to rewrite the numerator and denominator and four divided by itself is one.

$4\frac{1}{3}\times 5\frac{1}{2}\neq 20\frac{1}{6}$ The sixth grader has multiplied whole numbers and then multiplied fractions.

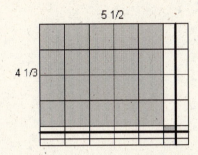

The shaded region above shows the result of the sixth grader and notice the difference between this image

and the following.

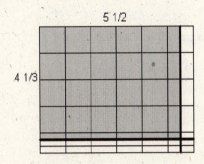

The correct product includes the five thirds horizontally as well as the four halves vertically, thus from

counting off of the image

$$4\frac{1}{3}\times 5\frac{1}{2}=20+\frac{1}{3}+\frac{1}{3}+\frac{1}{3}+\frac{1}{3}+\frac{1}{3}+\frac{1}{2}+\frac{1}{2}+\frac{1}{2}+\frac{1}{2}+\frac{1}{6}$$

$$=20+\frac{5}{3}+2+\frac{1}{6}$$

$$=22+1\frac{2}{3}+\frac{1}{6}$$

$$=23+\frac{4}{6}+\frac{1}{6}$$

$$=23\frac{5}{6}$$

LE7   Connection

a) $3 \div \frac{1}{4}$ Means "how many $\frac{1}{4}$s does it take to make three?" Here is a representation for three.

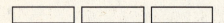

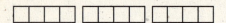

It takes four $\frac{1}{4}$s for each one, so a total of twelve $\frac{1}{4}$s to make three. Therefore $3 \div \frac{1}{4} = 12$

b) With a common denominator $3 \div \frac{1}{4} = \frac{12}{4} \div \frac{1}{4}$ Now $\frac{1}{4}$ goes into $\frac{12}{4}$ twelve times. Therefore $3 \div \frac{1}{4} = 12$

LE8   Connection

$\frac{3}{4} \div \frac{1}{8}$ Means how many $\frac{1}{8}$s make $\frac{3}{4}$? From the diagram below we can count to see that the answer is six.

Therefore $\frac{3}{4} \div \frac{1}{8} = 6$

LE9   Reasoning

a) In Example 3, where $4 \div \frac{1}{2} = 8$, dividing by $\frac{1}{2}$ is the same as multiplying by 2.

b) In LE 7, dividing by $\frac{1}{4}$ is the same as multiplying by 4.

c) In LE 8, dividing by $\frac{1}{8}$ is the same as multiplying by 8.

d) Parts (a) – (c) suggest that for rational numbers $\frac{a}{b}$ and $\frac{c}{d}$ with $\frac{c}{d} \neq 0$, $\frac{a}{b} \div \frac{c}{d} = \frac{a}{b} \times \frac{d}{c} = \frac{ad}{bc}$.

LE10   Skill

$$1\frac{2}{3} \div 1\frac{1}{2} = \frac{5}{3} \div \frac{3}{2}$$

$$= \frac{5}{3} \times \frac{2}{3}$$

$$= \frac{5 \times 2}{3 \times 3}$$

$$= \frac{10}{9}$$

$$= 1\frac{1}{9}$$

LE11   Concept

Any nonzero rational number multiplied by its reciprocal equals one. $\dfrac{a}{b} \times \dfrac{b}{a} = \dfrac{ab}{ba}$

$$= \frac{ab}{ab}$$

$$= 1$$

LE12   Skill

$$\frac{3}{5} \div \frac{2}{7} = \frac{\frac{3}{5}}{\frac{2}{7}}$$

$$= \frac{\frac{3}{5} \times \frac{7}{2}}{\frac{2}{7} \times \frac{7}{2}}$$

$$= \frac{\frac{3}{5} \times \frac{7}{2}}{1}$$

$$= \frac{3}{5} \times \frac{7}{2}$$

LE13   Connection

Division, equal measures

LE14   Connection

Multiplication, equal groups

LE15   Concept

a) You can make 2 groups of 4

b) There are 3 leftover

c) $\dfrac{3}{4}$

d) $2R3$ or $2\dfrac{3}{4}$

LE16   Reasoning

1) When $\dfrac{c}{d}$ is greater than one

2) When $\dfrac{c}{d}$ is equal to one

3) When $\dfrac{c}{d}$ is less than one

LE17   Concept

$\dfrac{3}{16}\times\dfrac{3}{5}$ is (c) less than $\dfrac{3}{16}$

LE18   Concept

$9\dfrac{1}{2}\div\dfrac{3}{4}$ is (a) greater than $9\dfrac{1}{2}$

LE19   Summary

To multiply fractions $\dfrac{a}{b}\times\dfrac{c}{d}$ where $b\neq 0$ and $d\neq 0$ we multiply numerators and multiply denominators,

$\dfrac{a\times c}{b\times d}$ then check to see if this product is reducible by a common factor.

To divide fractions $\dfrac{a}{b}\div\dfrac{c}{d}$ where $\dfrac{c}{d}\neq 0$ we invert and multiply the denominator $\dfrac{a}{b}\div\dfrac{c}{d}=\dfrac{a}{b}\times\dfrac{d}{c}$ and then

apply what we learned about multiplying fractions, $\dfrac{a\times d}{b\times c}$.

Homework Exercises 6.3

1. $4 \times \dfrac{1}{5} = \dfrac{1}{5} + \dfrac{1}{5} + \dfrac{1}{5} + \dfrac{1}{5} = \dfrac{4}{5}$

3. $\dfrac{1}{5} \times \dfrac{1}{3}$ Can be shown by these diagrams, begin with $\dfrac{1}{3}$

1/3

Now insert four horizontal lines

1/3

The result is the figure has been divided into 15 congruent pieces. Now darken $\dfrac{1}{5}$ of the shaded part,

 Which is $\dfrac{1}{5}$ of $\dfrac{1}{3}$ or $\dfrac{1}{15}$.

5. $\dfrac{3}{4} \times \dfrac{4}{5}$ Can be shown by these diagrams, begin with $\dfrac{4}{5}$

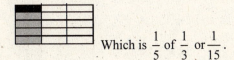

Now insert three horizontal lines

The result is the figure has been divided into 20 congruent pieces. Now darken $\dfrac{3}{4}$ of the shaded part,

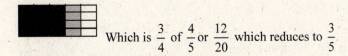 Which is $\dfrac{3}{4}$ of $\dfrac{4}{5}$ or $\dfrac{12}{20}$ which reduces to $\dfrac{3}{5}$

7. This grid shows the product of $\dfrac{4}{5}$ and $\dfrac{1}{4}$ which is $\dfrac{4}{20}$ which reduces to $\dfrac{1}{5}$

9.
a) $6 \times \frac{1}{3}$ Means six measures of $\frac{1}{3}$

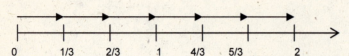

So $6 \times \frac{1}{3} = 2$

b) $\frac{1}{3} \times 6$ Means $\frac{1}{3}$ of 6

1/3

So $\frac{1}{3}$ of 6 is 2

11.
$$5\frac{1}{2} \times 2\frac{2}{3} = \frac{11}{2} \times \frac{8}{3} = \frac{11 \times 8}{2 \times 3}$$

This represents one fraction, you can divide the numerator and denominator by the common factor of two,

$$\frac{11 \times 8}{2 \times 3} = \frac{11 \times 4 \times 2}{2 \times 3}$$

$$= \frac{11 \times 4}{3}$$

13.
a) $2\frac{1}{2} \times 1\frac{1}{2}$ From the diagram we see 2 units, three $\frac{1}{2}$s, and one $\frac{1}{4}$ so $2\frac{1}{2} \times 1\frac{1}{2} = 2 + \frac{1}{2} + \frac{1}{2} + \frac{1}{2} + \frac{1}{4}$

$$= 2 + 1 + \frac{1}{2} + \frac{1}{4}$$

$$= 3\frac{3}{4}$$

b) $3\frac{1}{2} \times 2\frac{1}{2}$

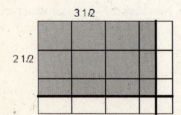

From the diagram we see 6 units, five $\frac{1}{2}$s, and one $\frac{1}{4}$ so $3\frac{1}{2}\times2\frac{1}{2}=6+\frac{1}{2}+\frac{1}{2}+\frac{1}{2}+\frac{1}{2}+\frac{1}{2}+\frac{1}{4}$

$$=6+2+\frac{1}{2}+\frac{1}{4}$$

$$=8\frac{3}{4}$$

c) $4\frac{1}{2}\times3\frac{1}{2}$

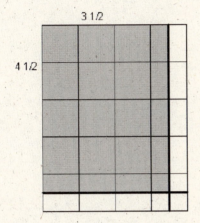

From the diagram we see 12 units, seven $\frac{1}{2}$s, and one $\frac{1}{4}$ so $4\frac{1}{2}\times3\frac{1}{2}=12+\frac{1}{2}+\frac{1}{2}+\frac{1}{2}+\frac{1}{2}+\frac{1}{2}+\frac{1}{2}+\frac{1}{2}+\frac{1}{4}$

$$=12+3+\frac{1}{2}+\frac{1}{4}$$

$$=15\frac{3}{4}$$

d)

$$a\frac{1}{2}\times b\frac{1}{2}=\left(a+\frac{1}{2}\right)\times\left(b+\frac{1}{2}\right)$$

$$=\left(b+1+\frac{1}{2}\right)\times\left(b+\frac{1}{2}\right)$$

$$=\left(b+\frac{3}{2}\right)\times\left(b+\frac{1}{2}\right)$$

$$=b^2+\frac{1}{2}b+\frac{3}{2}b+\frac{3}{4}$$

$$=b^2+2b+\frac{3}{4}$$

$$=b(b+2)+\frac{3}{4}$$

e)   If you use part d) as a shortcut you must recognize that commutative law of multiplication allows us to

rewrite the problem, $18\frac{1}{2} \times 19\frac{1}{2} = 19\frac{1}{2} \times 18\frac{1}{2}$, this is important as now $a = b+1$ that is to say $19 = 18+1$.

$$19\frac{1}{2} \times 18\frac{1}{2} = 18(18+2) + \frac{3}{4}$$

$$= 18(20) + \frac{3}{4}$$

$$= 360\frac{3}{4}$$

15. $3\frac{1}{2} \times 4\frac{1}{8}$

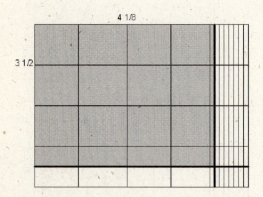

From the model we can count to find the product, there are 12 units, three $\frac{1}{8}$ s, four $\frac{1}{2}$ s, and one $\frac{1}{16}$ so,

$$3\frac{1}{2} \times 4\frac{1}{8} = 12 + \frac{1}{8} + \frac{1}{8} + \frac{1}{8} + \frac{1}{2} + \frac{1}{2} + \frac{1}{2} + \frac{1}{2} + \frac{1}{16}$$

$$= 12 + \frac{3}{8} + 2 + \frac{1}{16}$$

$$= 14 + \frac{6}{16} + \frac{1}{16}$$

$$= 14\frac{7}{16}$$

17. For student

19. a) $\dfrac{1}{2}$ of 8 is represented using multiplication as $\dfrac{1}{2} \times 8$

b) $\dfrac{1}{2}$ of 8 is represented using division as $8 \div 2$

21. a) $\dfrac{2}{3} \times \dfrac{3}{4} = \dfrac{2 \times 3}{3 \times 4}$

$$= \dfrac{2 \times 3}{3 \times 2 \times 2}$$

$$= \dfrac{\cancel{2} \times \cancel{3}}{3 \times \cancel{2} \times 2}$$

$$= \dfrac{1}{2}$$

b) $2\dfrac{1}{3} \times \dfrac{2}{3} = \dfrac{7}{3} \times \dfrac{2}{3}$

$$= \dfrac{7 \times 2}{3 \times 3}$$

$$= \dfrac{14}{9}$$

$$= 1\dfrac{5}{9}$$

23. a) Using a measurement picture $2 \div \frac{1}{4}$ can be thought of as how many $\frac{1}{4}$s are in 2?

 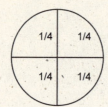

From the picture we see there are eight $\frac{1}{4}$s in 2.

b) To compute $2 \div \frac{1}{4}$ with a common denominator first rewrite 2 with a denominator of 4.

$2 = \frac{2}{1} \times \frac{4}{4}$ Now use this $\frac{8}{4} \div \frac{1}{4}$ and ask the question, "How many quarters in eight quarters?" The answer is

$= \frac{8}{4}$

8, therefore $\frac{8}{4} \div \frac{1}{4} = 8$

25. $\frac{3}{4} \div \frac{1}{8}$ Means how many $\frac{3}{4}$s make $\frac{1}{2}$. The whole is the complete circle, divide the circle in half then divide

the circle into eighths to see what part of the whole the problem refers to.

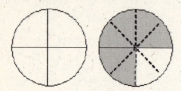

The dashed line segments represent the eighths and from the diagram we see that six $\frac{1}{8}$s make $\frac{3}{4}$ or

$\frac{3}{4} \div \frac{1}{8} = 6$

27. $\frac{3}{4} \div 2$ Means divide the $\frac{3}{4}$ into two equal parts

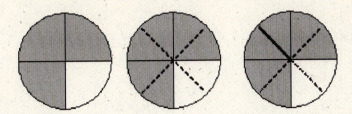

First we show $\frac{3}{4}$ then split each quarter in half and the solid segment on the right hand image shows where

to divide $\frac{3}{4}$. Notice when $\frac{3}{4}$ is divided in two there are $\frac{3}{8}$ on either side of the solid segment, therefore

$$\frac{3}{4} \div 2 = \frac{3}{8}$$

29. $1\frac{1}{4} \div \frac{1}{2}$ Means how many $\frac{1}{2}$s make $1\frac{1}{4}$?

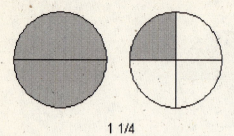

1 1/4

From the picture we see there are two $\frac{1}{2}$s and half of a $\frac{1}{2}$, so $1\frac{1}{4} \div \frac{1}{2} = 2\frac{1}{2}$

31. a) $\frac{3}{4}$ of the whole is shaded

b) The whole is $\frac{4}{3}$ of the shaded section

c) One shaded part of the three shaded parts shows $\frac{1}{3}$ of $\frac{3}{4}$

d) $1 \div \frac{3}{4}$ is shown by the shaded sections in the whole

33.
$$\frac{10}{21} \div \frac{2}{3} = \frac{n}{m}$$

$$\frac{10}{21} \times \frac{3}{2} = \frac{n}{m}$$

$$\frac{10 \times 3}{21 \times 2} = \frac{n}{m}$$

$$\frac{2 \times 5 \times 3}{7 \times 3 \times 2} = \frac{n}{m}$$

$$\frac{5}{7} = \frac{n}{m}$$

35. a) $\frac{5}{8} \div \frac{1}{2} = \frac{5}{8} \times \frac{2}{1}$

$$= \frac{5 \times 2}{8 \times 1}$$

$$= \frac{5 \times 2}{4 \times 2 \times 1}$$

$$= \frac{5}{4} \text{ or } 1\frac{1}{4}$$

b) $-8 \div 3\frac{1}{9} = \frac{-8}{1} \div \frac{28}{9}$

$$= \frac{-8}{1} \times \frac{9}{28}$$

$$= \frac{-8 \times 9}{1 \times 28}$$

$$= \frac{-4 \times 2 \times 9}{1 \times 4 \times 7}$$

$$= \frac{-18}{7} \text{ or } -2\frac{4}{7}$$

c) $\frac{x}{5} \div \frac{x}{7} = \frac{x}{5} \times \frac{7}{x}; \ x \neq 0$

$$= \frac{7x}{5x}$$

$$= \frac{7}{5} \text{ or } 1\frac{2}{5}$$

37.   a) $\frac{1}{2}x = 10$, Ask yourself, "Half of what is ten?" Half of twenty is ten.

b) $4 \div \frac{1}{2} = y$, Ask yourself "How many halves in four?" There are two halves in one, so there are eight

halves in four.

39. $\frac{3}{8} \times \frac{1}{6} = \frac{3 \times 1}{8 \times 6}$

$= \frac{3 \times 1}{8 \times 2 \times 3}$

$= \frac{1}{16}$

41.

$$\frac{1}{2} \div \frac{3}{4} = \frac{\frac{1}{2}}{\frac{3}{4}}$$

$$= \frac{\frac{1}{2} \times \frac{4}{3}}{\frac{3}{4} \times \frac{4}{3}}$$

$$= \frac{\frac{1}{2} \times \frac{4}{3}}{1}$$

$$= \frac{1}{2} \times \frac{4}{3}$$

43.

$$\frac{a}{b} \div \frac{c}{d} = \frac{\frac{a}{b}}{\frac{c}{d}}$$

$$= \frac{\frac{a}{b} \times \frac{d}{c}}{\frac{c}{d} \times \frac{d}{c}}$$

$$= \frac{\frac{a}{b} \times \frac{d}{c}}{1}$$

$$= \frac{a}{b} \times \frac{d}{c}$$

$$= \frac{ad}{bc}$$

45. You could present the problems b, c, a, d. Choice b is first since the 3 is a nice denominator relative to 15, same reason for choice c, choice a is next as multiplying 2 and 4 is a little nicer than the product of 4 and 6 in choice d.

47. a)

| 1/4 | 1/4 | 1/4 | 1/4 | 1/4 | 1/4 | =1 ½ cups

b) $\dfrac{1}{4}+\dfrac{1}{4}+\dfrac{1}{4}+\dfrac{1}{4}+\dfrac{1}{4}+\dfrac{1}{4}=\dfrac{6}{4}$ OR $6\times\dfrac{1}{4}=\dfrac{6}{1}\times\dfrac{1}{4}$

$$=1\dfrac{2}{4} \qquad\qquad =\dfrac{6}{4}$$

$$=1\dfrac{1}{2} \qquad\qquad =1\dfrac{2}{4}$$

$$=1\dfrac{1}{2}$$

c) Part a) illustrates multiplication of equal measures.

49. a) The wall is $82\dfrac{1}{2}$ inches high and the tiles are $5\dfrac{1}{2}$ inch squares, so there are

$$82\dfrac{1}{2}\div 5\dfrac{1}{2}=\dfrac{165}{2}\div\dfrac{11}{2}$$

$$=\dfrac{165}{2}\times\dfrac{2}{11}$$

$$=\dfrac{165}{11}$$

$$=15 \text{ tiles}$$

b) Part a) represents division and the category is equal measures

51. a) Four cartons are ten feet high, so each carton is $\dfrac{10}{4}=2\dfrac{1}{2}$ feet high. There is eight feet remaining until the

ceiling, so $8\div 2\dfrac{1}{2}=8\div\dfrac{5}{2}$

$$=\dfrac{8}{1}\times\dfrac{2}{5}$$

$$=\dfrac{16}{5} \text{ or } 3\dfrac{1}{5}$$

Since we don't have partial cartons we can stack three more cartons before we reach the ceiling.

53. $\frac{1}{4}$ of $\frac{4}{5}$ brought in canned soup,

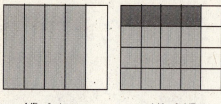

4/5 of class   1/4 of 4/5

Notice that the picture on the right is twenty equal parts and $\frac{1}{4}$ of $\frac{4}{5}$ is $\frac{4}{20}$ or $\frac{1}{5}$

55. a

57.

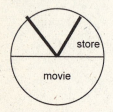

The whole circle can separated into six equal parts, two of these equal parts represents the $4 Carol has left, so each one sixth section represents $2. Therefore Carol began the day with $12.

59.   a) I have $\frac{4}{5}$ of a cup of milk in the refrigerator. I need $\frac{2}{3}$ of the $\frac{4}{5}$ cup for a recipe. How much milk do I

need for the recipe?

b) How many $\frac{1}{4}$ s are in $\frac{5}{8}$?

61.   If I had $10 and divided it in half I would have $5. If I had $10 and divided it by $0.50, I would have 20

fifty cent pieces.

63.   a)

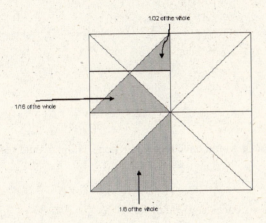

$$\frac{1}{32}+\frac{1}{16}+\frac{1}{8}=\frac{1}{32}+\frac{2}{32}+\frac{4}{32}$$
$$=\frac{7}{32}$$

b) Subdivide the whole into right triangles that represent $\frac{1}{32}$ of the entire whole. The shaded pieces would

represent $\frac{7}{32}$ of the whole.

65.   a) If Ward borrows another hog then the total is 18. So each child will receive:

Wacky; $\frac{4}{9}\times18=8$ hogs

Harpo; $\frac{1}{3}\times18=6$ hogs

Loopy II; $\frac{1}{6}\times18=3$ hogs

b) Wiggy's approach seems to work since $\dfrac{4}{9}+\dfrac{1}{3}+\dfrac{1}{6}=\dfrac{8}{18}+\dfrac{6}{18}+\dfrac{3}{18}$ and $\dfrac{17}{18}$ of 18 is 17.

$$=\dfrac{17}{18}$$

c) Wacky; $\dfrac{4}{9}\times 17=\dfrac{68}{9}$ hogs

$$=7\dfrac{5}{9}$$

Harpo; $\dfrac{1}{3}\times 17=\dfrac{17}{3}$ hogs

$$=6\dfrac{1}{3}$$

Loopy II; $\dfrac{1}{6}\times 17=\dfrac{17}{6}$ hogs

$$=2\dfrac{5}{6}$$

If the will instructed there to be rounding, then Wacky would have received 8 hogs, Harpo would receive 6 hogs and Loopy II would receive 3 hogs.

67.   a) You have 46 cans of juice. How many complete six packs can you make?

b) You have 46 cans of juice that you want to pack in cartons of six. How many cartons are needed to pack all the cans?

c) Put the 46 cans of juice in six packs. How many cans are left over?

d) How many cartons of juice do you have if you pack the 46 cans in cartons of six?

69.   a) Both fractions have the same numerator. The denominator of the second fraction equals the sum of the numerator and denominator from the first fraction.

b) Example one; $\dfrac{3}{5}-\dfrac{3}{8}=\dfrac{24}{40}-\dfrac{15}{40}$, $\dfrac{3}{5}\times\dfrac{3}{8}=\dfrac{9}{40}$

$$=\dfrac{9}{40}$$

Example two; $\dfrac{1}{2}-\dfrac{1}{3}=\dfrac{3}{6}-\dfrac{2}{6}$, $\dfrac{1}{2}\times\dfrac{1}{3}=\dfrac{1}{6}$

$$=\dfrac{1}{6}$$

71.   For student

Lesson Exercises 6.4

LE1   Opener

a) Addition is commutative for rational numbers

b) Multiplication is associative for rational numbers

c) Multiplication is distributive over addition for rational numbers

d) (a) – (c) conclusions are based on induction

LE2   Concept

The additive inverse of $\dfrac{3}{4}$ is $-\dfrac{3}{4}$

LE3   Concept

Associative property of addition guarantees that $\left(x + \dfrac{1}{2} \right) + 2 = x + \left(\dfrac{1}{2} + 2 \right)$

LE4   Skill

$$8 \times 5\frac{1}{2} = 8\left(5 + \frac{1}{2}\right)$$
$$= 8(5) + 8\left(\frac{1}{2}\right)$$
$$= 40 + 4$$
$$= 44$$

LE5   Reasoning

The counter examples for whole number and integer subtraction and division would also be counter examples for rational numbers. If a rule does not apply to all whole numbers, then it cannot apply to all rational numbers, since every whole number is also a rational number.

LE6   Skill

a) The multiplicative inverse of -4 is $-\dfrac{1}{4}$

b) The multiplicative inverse of $\dfrac{2}{3}$ is $\dfrac{3}{2}$

LE7   Reasoning

a) Zero can be written as a rational number but its multiplicative inverse is undefined

b) Zero is the only rational number without a multiplicative inverse

LE8   Skill

a) You can use rounding for the amount of tomato sauce needed by rounding $4\dfrac{1}{2}$ to 5 and rounding $2\dfrac{3}{4}$ to

3, so Celia needs approximately $5 \times 3 = 15$ cups

b) The estimate of 15 cups is too high

c) By rounding $4\dfrac{1}{2}$ to 4 and rounding $2\dfrac{3}{4}$ to 2, Celia needs approximately $4 \times 2 = 8$ cups, so the range of

needed tomato sauce is 8 to 15 cups

LE9   Skill

Round $\dfrac{5}{12}$ to $\dfrac{1}{2}$ and round $\dfrac{7}{8}$ to 1 so $\dfrac{5}{12} + \dfrac{7}{8} \approx \dfrac{1}{2} + 1$ or $\dfrac{3}{2}$

LE10   Skill

You can estimate the amount of the discount by rounding $\dfrac{3}{5}$ to $\dfrac{1}{2}$, and \$604 to \$600, half of \$600 is \$300

LE11   Reasoning

a) $\dfrac{1}{5} + \dfrac{3}{4} = \dfrac{4}{9}$ $\dfrac{1}{2} + \dfrac{2}{3} = \dfrac{3}{5}$

b) The error pattern is the addition of numerators and the addition of denominators without regard for

common denominators

c) $\dfrac{1}{5} + \dfrac{3}{4} = \dfrac{4}{9}$ The sum $\dfrac{4}{9}$ is smaller than the $\dfrac{3}{4}$ addend so how can $\dfrac{1}{5} + \dfrac{3}{4} = \dfrac{4}{9}$?

LE12   Reasoning

a) $\dfrac{4}{7} \div \dfrac{2}{3} = \dfrac{14}{12}$ $\dfrac{7}{8} \div 2 = \dfrac{16}{7}$

b) The child is inverting the first fraction, changing the operation to multiplication, and multiplying the "new" fraction by the second fraction.

c) $8\dfrac{1}{2} \div 2\dfrac{2}{3} \approx 9 \div 3$ But the quotient given by this error pattern, $\dfrac{16}{51}$ is less than $\dfrac{1}{2}$, how can that be?

LE13   Reasoning

a) $\dfrac{12}{3} = 4$ $\dfrac{1}{4} = 4$

b) The child divides the smaller number into the larger number

c) Estimation used correctly would reveal to the child that a denominator larger than the numerator is a value less than one.

LE14   Reasoning

a)
$\begin{array}{ll} 5\dfrac{1}{4} = \dfrac{2}{8} & 6\dfrac{1}{6} = \dfrac{2}{12} \\ +2\dfrac{1}{2} = \dfrac{4}{8} & +4\dfrac{1}{4} = \dfrac{3}{12} \\ \hline \quad\quad\; \dfrac{6}{8} & \quad\quad\; \dfrac{5}{12} \end{array}$

b) The child is finding a common denominator for the fractional part, summing the fractions and ignoring the addition of the whole numbers

c)
$\begin{array}{l} 5\dfrac{1}{4} \approx 5 \\ +2\dfrac{1}{2} \approx 3 \\ \hline \quad\quad\; 8 \end{array}$ The child's answer of $\dfrac{6}{8}$ is less than one.

LE15   Summary

All of the properties for integers are retained for rational numbers. Rational numbers, except zero, have multiplicative inverses.

1. a) Addition and multiplication are commutative for rational numbers

    b) Addition and multiplication are associative for rational numbers

3. According to the associative property of multiplication for a rational number x, $\frac{1}{2} \bullet (4 \bullet x) = \left(\frac{1}{2} \bullet 4\right) \bullet x$

5. The additive inverse of $-2\frac{1}{2}$ is $2\frac{1}{2}$

7. Mentally the total distance traveled can be calculated using the distributive property of multiplication over addition.

$$50 \times 6\frac{1}{2} = 50\left(6 + \frac{1}{2}\right)$$
$$= (50 \bullet 6) + \left(50 \bullet \frac{1}{2}\right)$$
$$= 300 + 25$$
$$= 325 \text{ miles}$$

9. $\frac{3}{4} \bullet 8\frac{1}{2} = \frac{3}{4}\left(8 + \frac{1}{2}\right)$ is a correct use of the distributive property

11.

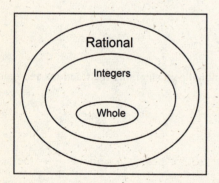

13. A counter example for whole numbers, $3 - 2 \neq 2 - 3$, would also be a counter example for integers and rational numbers

15. The multiplicative inverse of $-2\frac{1}{2}$ is $-\frac{2}{5}$. To find the inverse rewrite $-2\frac{1}{2}$ as $-\frac{5}{2}$

17. a) Rational numbers have the denseness property, whole numbers and integers do not have this property.

   b) All nonzero rational numbers have the property of multiplicative inverses, whole numbers and integers

   do not have this property.

19. Round $1\frac{3}{4}$ to 2 cups and round $8\frac{1}{4}$ to 8 pounds, so you need approximately 2 cups of sugar for every pound

   of belly busters, so for 8 pounds you would need approximately 16 cups of sugar.

21. $26\frac{7}{3600} \times 32\frac{9}{60} \approx 30 \times 30$ Or about 900, answer (d)

23. a) Round $\frac{7}{8}$ to 1 and $\frac{1}{16}$ to 0, so the two boards nailed together would be approximately 1" thick

   b) These solutions by children could be the result of adding numerators, $7+1=8$ or adding denominators

   $8+16=24$ without regard to finding common denominators then finding the sum.

25. a) $\frac{2}{9}$ is close to $\frac{1}{4}$

   b) $\frac{18}{23}$ is close to $\frac{3}{4}$

   c) $\frac{5}{11}$ is close to but less than $\frac{1}{2}$

   d) $\frac{2}{35}$ is close to 0

   e) $\frac{9}{17}$ is close to but greater than $\frac{1}{2}$

27. A simple fraction that approximates the fraction of children that bring lunch to school is $\frac{41}{126}$, the

   numerator is approximately 40 and the denominator is approximately 120, so a reasonable approximation

   could be $\frac{40}{120} = \frac{1}{3}$

29. For this class of fifth graders 23 is approximately 24 and one fourth of twenty four is $\frac{1}{4} \cdot 24 = 6$ students

   whom like vanilla ice cream

31. Consider the $17\frac{1}{4}$ panels as approximately 20 inches and the wall width $384\frac{1}{2}$ as approximately 400 inches, so you will need about $400 \div 20 = 20$ panels

33. $872\frac{7}{16} \div \frac{3}{8}$ is approximately $1000 \div \frac{1}{2} = 1000 \bullet 2$
$$= 2000$$

choice (d)

35. a) $\dfrac{1}{2} + \dfrac{3}{4} = \dfrac{2+3}{4+4} = \dfrac{5}{8}$

$\dfrac{2}{3} + \dfrac{1}{6} = \dfrac{4+1}{6+6} = \dfrac{5}{12}$

$\dfrac{1}{5} + \dfrac{9}{10} = \dfrac{2+9}{10+10} = \dfrac{11}{20}$

b) After obtaining the common denominator the child incorrectly adds the numerators and the denominators

37. a) $4\dfrac{1}{5} = 3\dfrac{11}{5}$

$\underline{-2\dfrac{2}{5} = -2\dfrac{2}{5}}$

$\qquad\quad 1\dfrac{9}{5} = 2\dfrac{4}{5}$

$3\dfrac{1}{3} = 3\dfrac{11}{3}$

$\underline{-1\dfrac{2}{3} = -1\dfrac{2}{3}}$

$\qquad\quad 2\dfrac{9}{3} = 5$

$8\dfrac{2}{5} = 8\dfrac{12}{5}$

$\underline{-3\dfrac{3}{5} = -3\dfrac{3}{5}}$

$\qquad\quad 5\dfrac{9}{5} = 6\dfrac{4}{5}$

b) The child adds ten to the numerator of the fraction in their regrouping strategy

39. a) $\dfrac{4+\cancel{3}}{2+\cancel{3}}=\dfrac{4}{2}=2$

$\dfrac{5+2}{5+3}=\dfrac{\cancel{5}+2}{\cancel{5}+3}=\dfrac{2}{3}$

$\dfrac{y+3}{y+1}=\dfrac{\cancel{y}+3}{\cancel{y}+1}=\dfrac{3}{1}=3$

b) The child slashes common addends rather than common factors in the numerator and the denominator, yielding the effect of a quotient of one

41. a) $1=\dfrac{2}{1+1}=\dfrac{2}{1}+\dfrac{2}{1}=4$ The error occurs in the claim that $\dfrac{2}{1+1}=\dfrac{2}{1}+\dfrac{2}{1}$, this is not true

b) $10=\dfrac{20}{2}=\dfrac{10+10}{2}=\dfrac{10}{2}+10=15$ The error occurs in the claim that $\dfrac{10+10}{2}=\dfrac{10}{2}+10$ this is not true

43. $\left(\dfrac{u}{v}\bullet\dfrac{w}{x}\right)\bullet\dfrac{y}{z}=\dfrac{uw}{vx}\bullet\dfrac{y}{z}=\dfrac{uwy}{vxz}$ And $\dfrac{u}{v}\bullet\left(\dfrac{w}{x}\bullet\dfrac{y}{z}\right)=\dfrac{u}{v}\bullet\dfrac{wy}{xz}=\dfrac{uwy}{vxz}$ equivalent statements therefore the

multiplication of rational numbers is associative

45. a) $\dfrac{2}{x}+\dfrac{2}{y}=\dfrac{2}{x+y}$ Is false, a counterexample is Let $x=1$ and $y=1$

$$\dfrac{2}{1}+\dfrac{2}{1}\overset{?}{=}\dfrac{2}{1+1}$$

$$2+2\overset{?}{=}\dfrac{2}{2}$$

$$4\neq1$$

b) $\dfrac{x+y}{y}=x$ Is false, a counterexample is Let $x=1$ and $y=1$

$$\dfrac{1+1}{1}\overset{?}{=}1$$

$$\dfrac{2}{1}\overset{?}{=}1$$

$$2\neq1$$

Chapter 6 Review Exercises

1. $\frac{5}{6}$ Represents

Part of a Whole

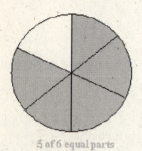

5 of 6 equal parts

Part of a set

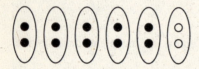

5 of 6 equal groups

Division

$5 \div 6$

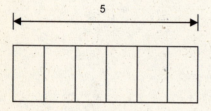

Divide 5 into 6 equal parts

Number line location

$\frac{5}{6}$ of the way from 0 to 1

3. a) $\frac{1}{6}$ is represented as

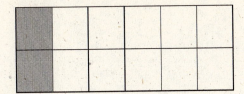

b) If $\frac{1}{2}$ is represented by

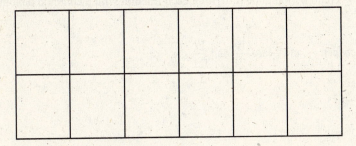

Then one can be represented as

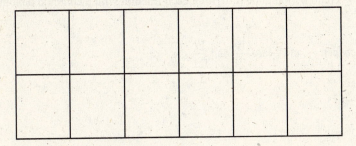

So $\frac{3}{4}$ is represented as

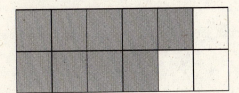

5. $\frac{1}{2}$ is a rational number but not an integer

7. $135 = 3^3 \cdot 5$

$162 = 2 \cdot 3^4$

Therefore

$GCF(135,162) = 3^3$

Using the GCF to simplify $\dfrac{135}{162} = \dfrac{3^3 \cdot 5}{3^3 \cdot 3 \cdot 2}$

$= \dfrac{5}{3 \cdot 2}$

$= \dfrac{5}{6}$

9.   Mathematics Professors $\dfrac{5}{8} \cdot \dfrac{3}{3} = \dfrac{15}{24}$

Science Professors $\dfrac{4}{6} \cdot \dfrac{4}{4} = \dfrac{16}{24}$

$\dfrac{16}{24} > \dfrac{15}{24}$ so the Science faculty get slightly larger portions.

11.   a) Two fractions between $\dfrac{5}{7}$ and $\dfrac{6}{7}$ could be found be re-writing each fraction with a common denominator

of 21, $\dfrac{5}{7} = \dfrac{15}{21}$ so two fractions between $\dfrac{5}{7}$ and $\dfrac{6}{7}$ are $\dfrac{16}{21}$ and $\dfrac{17}{21}$.

$\dfrac{6}{7} = \dfrac{18}{21}$

13.   a) Fraction bars for $\dfrac{2}{3} + \dfrac{1}{6}$

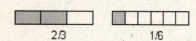

2/3   1/6

b) A common denominator is needed for a fraction bar that shows the sum as the illustration above shows,

we cannot mix thirds with sixths.

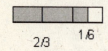

2/3   1/6

c)

2/3 + 1/6 = 5/6

15. a) $\dfrac{3}{8}+\dfrac{2}{9}=\dfrac{27}{72}+\dfrac{16}{72}$

$=\dfrac{43}{72}$

b) $42\dfrac{5}{12}-24\dfrac{2}{3}=41\dfrac{17}{12}-24\dfrac{8}{12}$

$=17\dfrac{9}{12}$

$=17\dfrac{3}{4}$

c) $-3\dfrac{1}{2}-2\dfrac{1}{4}=-3\dfrac{1}{2}+-2\dfrac{1}{4}$

$=-3\dfrac{2}{4}+-2\dfrac{1}{4}$

$=-5\dfrac{3}{4}$

17. $\dfrac{1}{4}\times\dfrac{8}{9}=\dfrac{1\times 8}{4\times 9}$ Now that this is all one fraction, you can divide the numerator and the denominator by a

common factor of four, then $\dfrac{8}{4}$ becomes $\dfrac{2}{1}$. $\dfrac{1}{\cancel{4}}\times\dfrac{\cancel{8}^{2}}{9}=\dfrac{2}{9}$

19. a) $4 \div \dfrac{1}{3}$

With a picture, first represent four then divide four into thirds.

There are twelve "one thirds" that make four.

b) $4 \div \dfrac{1}{3}$ with a common denominator

$$4 \div \frac{1}{3} = \frac{4}{1} \div \frac{1}{3}$$
$$= \frac{12}{3} \div \frac{1}{3}$$
$$= \frac{12}{3} \times \frac{3}{1}$$
$$= 4$$

21.

$$\frac{2}{5} \div \frac{7}{9} = \frac{\dfrac{2}{5}}{\dfrac{7}{9}}$$
$$= \frac{\dfrac{2}{5} \times \dfrac{9}{7}}{\dfrac{9}{7}}$$
$$= \frac{\dfrac{2}{5} \times \dfrac{9}{7}}{1}$$
$$= \frac{2}{5} \times \frac{9}{7}$$

23. a) $\frac{1}{10}+\frac{1}{3}=\frac{1}{10}\cdot\frac{3}{3}+\frac{1}{3}\cdot\frac{10}{10}$ So $\frac{13}{30}$ of your income goes out to charities and taxes, leaving $\frac{30}{30}-\frac{13}{30}=\frac{17}{30}$ of

$$=\frac{3}{30}+\frac{10}{30}$$

$$=\frac{13}{30}$$

your income for other expenses.

b) Part a represents addition and combine groups, measures and subtraction, take away measures and

groups

25. A realistic word problem for $4\div\frac{1}{2}$ could be; I have four yards of fabric. A basket liner calls for one half of

a yard of fabric. How many basket liners can I make?

27. An example of the associative property of multiplication for rational numbers could be

$\left(\frac{1}{2}\times\frac{1}{3}\right)\times\frac{1}{4}=\frac{1}{2}\times\left(\frac{1}{3}\times\frac{1}{4}\right)$ Working out the right hand side and comparing the result to the arithmetic of the

left hand side reveals that associative property holds.

The left hand side: $\left(\frac{1}{2}\times\frac{1}{3}\right)\times\frac{1}{4}=\frac{1}{6}\times\frac{1}{4}$ The right hand side: $\frac{1}{2}\times\left(\frac{1}{3}\times\frac{1}{4}\right)=\frac{1}{2}\times\frac{1}{12}$

$$=\frac{1}{24}$$ $$=\frac{1}{24}$$

29. $24\times2\frac{1}{2}=(24\times2)+\left(24\times\frac{1}{2}\right)$

$$=48+12$$

$$=60$$

31. $472\frac{1}{4}\div\frac{3}{16}$ can be approximated by considering that $472\frac{1}{4}\approx500$ and $\frac{3}{16}\approx\frac{1}{4}$. So this problem reduces to

$500\div\frac{1}{4}=500\times4$ and the closest answer is choice a) 2500.

$$=2000$$

Chapter Seven

Section 7.1

LE 1   Opener

0.1 and 0.01 are not the same, the zero is a place holder, show the child a decimal square representation.

LE 2   Concept

a) In our numeration system, each place you move to the left multiplies place value by <u>ten</u>.

b) Each place you move to the right divides place value by <u>ten</u>, which is the same as multiplying by $\dfrac{1}{10}$.

LE 3   Connection

a) Some children believe that $0.32 > 0.40$ because they apply their rules for whole numbers that is to say

$32 > 4$.

b) Since 0.4 is four – tenths and 0.32 is thirty – two hundredths, use a decimal square picture to show that

$0.40 > 0.32$.

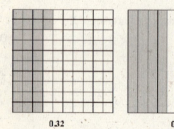

c) With a number line:

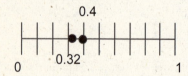

d)

Ones	Decimal	Tenths	Hundredths
0	.	3	2
0	.	4	0

LE 4   Skill

a) 0.621371 rounded to the preceding thousandth is 0.621

b) .621371 rounded to the next thousandth is 0.622

c) .621371 rounded to the nearest thousandth is 0.621

LE 5   Skill

a) When the child changes the problem to $9 \times 8 = \$72$ they understand rounding

b) When the child uses their calculator and estimates their answer to be $67 based on a calculator product

of 66.994 they understand rounding.

c) When the child changes the problem to $8 \times 7 = \$56$ they understand front end estimation.

d) When the child rounds 8.6 down to 8 and 7.79 up to 8 they understand rounding.

LE 6   Skill

The cost per pound of crabmeat could be estimated by compatible numbers, round the cost to $56 and the

quantity to 7 pounds, thus the cost per pound would be approximately $7\overline{)\$56}$ 8 $8 per pound.

LE 7   Reasoning

a) By extending a pattern
$$10^3 = 1000$$
$$10^2 = 100$$
$$10^1 = 10$$
$$10^0 = 1$$
$$10^{-1} = \frac{1}{10}$$
$$10^{-2} = \frac{1}{100}$$

b) Explain to the student that as the exponent decreases by one on the left the value on the right is the result

of dividing by ten each time.

LE 8   Concept

a) $(10^3) \bullet (10^4) = (10 \bullet 10 \bullet 10) \bullet (10 \bullet 10 \bullet 10 \bullet 10) = 10^7$

b) The short cut for multiplying $(10^3) \bullet (10^4)$ is to add the exponents and keep the base,

$(10^3) \bullet (10^4) = 10^{3+4}$

c) $10^3 \bullet 10^0 = 10^{3+0} = 10^3$

d) If $10^3 \bullet 10^0 = 10^3$ then 10^0 must equal one.

LE 9   Reasoning
a) When you multiply by 10^n, the decimal point moves n places to the right.

b) Inductive reasoning was used by observing the pattern and guessing the rule.

LE 10   Skill

To compute 4.6×10^3 mentally move the decimal point three places to the right, 4600.

LE 11   Skill

a) $150.5 billion is $150,500,000,000

b) The average cost per driver would be $\dfrac{150,500,000,000}{185,000,000} = \$813.51$

LE 12   Reasoning

a) On the basis of the given example, dividing a decimal number by 10^n, in which $n = 1,2,3,...$ appears to

be the same thing as moving the decimal point n places to the <u>left.</u>

b) The generalization in part (d) is formed with <u>inductive</u> reasoning.

LE 13   Skill

a) To calculate the exact answer, shift the decimal point three places to the left for $75, so each candy bar

costs $.075 or 7.5 cents.

b) Part (a) illustrates partition groups/measures

LE 14   Reasoning

A short cut you can use for multiplying by negative integer powers of ten, $a \times 10^c$ where c is a negative

integer, move the decimal point to the left c places.

LE 15   Skill

$$5.67 \times 10^{-6} = .00000567$$

LE 16   Opener

a)
$$\begin{array}{r} 400,000 \\ \times 360,000 \\ \hline 24000000000 \\ 1200000 \quad\quad \\ \hline 144,000,000,000 \end{array}$$

b)

```
400000*360000
            1.44E11
■
```

LE 17   Skill

0.00036 mile per hour in scientific notation is 3.6×10^{-4}

LE 18   Summary

Decimal squares are a good tool for comparing decimals. When you are multiplying by 10^n where

$n > 0$ you move the decimal point n places to the right. When $n < 0$, you move n places to the left.

7.1 Homework Exercises

1.   a) Forty – one and sixteen hundredths is 41.16

      b) Seven and five thousandths is 7.005

3.   If you don't see a decimal point in a number, it is understood to be to the right of the right most digit.

5.   a) The sign should say that apples are selling for $0.89/pound or 89¢/pound.

      b) If the store manager insists that the price is right absolutely buy the apples at that price.

7.   a) Some children think $0.11 > 0.2$ because they apply their whole number thinking, 11 is greater than 2.

b) $0.11 < 0.2$

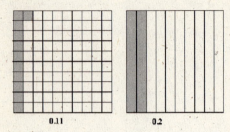

0.11   0.2

c)

d)

Ones	Decimal	Tenths	Hundredths
0	.	1	1
0	.	2	0

9.

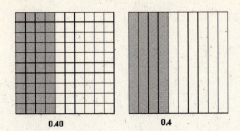

0.40   0.4

11.   a) $0.86 rounded to the next tenth is $0.9

b) $0.86 rounded to the preceding tenth is $0.8

c) $0.86 rounded to the nearest tenth is $0.9

13.   a) If labels are sold in packs of one hundred and you needed 640, you would have to purchase seven packs.

b) This application requires rounding up.

15.   Mount Everest is 1715.5 meters higher than Mount Api.

a) Using rounding the approximation is $8847.6 - 7132.1 \approx 8800 - 7100 = 1700$ m

b) Using the front end strategy $8847.6 - 7132.1 \approx 8000 - 7000 = 1000$ m

17.   The cost per ounce can be approximated by $\$1.29 \div 46 \approx \$1.20 \div 40 = \$0.03$

19.   The product is about $4 \times 40 = 160$, so the decimal point goes between the zero and the one, 160.11875

21. $10^3 = 1000$
$10^2 = 100$
$10^1 = 10$
$10^0 = 1$

23.   a) Each result equals the previous result divided by five.

b) Continuing this pattern $5^0 = 1$, $5^{-1} = \dfrac{1}{5}$, $5^{-2} = \dfrac{1}{25}$

25.   a) $\dfrac{2^6}{2^4} = \dfrac{2 \bullet 2 \bullet 2 \bullet 2 \bullet 2 \bullet 2}{2 \bullet 2 \bullet 2 \bullet 2} = 2^2$

b) The shortcut is $\dfrac{2^6}{2^4} = 2^{6-4} = 2^2$

c) $\dfrac{10^7}{10^4} = 10^{7-4} = 10^3$

d) $\dfrac{5^7}{5^{-3}} = 5^{7-(-3)} = 5^{10}$

e) $\dfrac{X^6}{X^2} = X^{6-2} = X^4$

27.   a) $3.62 \times 10^7 = 36,200,000$

b) $0.056 \times 10^5 = 5600$

29. $6.5 trillion is $6,500,000,000,000

31.   a) $1.2 \times 10^7 \text{ kg} = 12,000,000 \text{ kg}$

b) This is twelve million kilograms of dust.

33.   a) $4,268 \div 10^6 = 0.004268$

b) $3.62 \div 10^3 = 0.00362$

35.   The total cost computed mentally is to shift the decimal in $3.29 two places to the right, so the total cost is

$329.

37.   In standard decimal form $5.4 \times 10^{-6} = 0.0000054$

39. $3.4 \text{ E12} = 3.4 \times 10^{12}$

41.   a) $0.000000000000000000000013 = 1.3 \times 10^{-23}$

b) Scientific notation is more efficient when working with this small number.

c) On a calculator display this number could appear as:

```
1.3*10^(-23)
          1.3E-23
```

43. The U.S. population discards approximately $6 \times 291,000,000 = 1.746 \times 10^9$ pounds of trash per day.

45. This is not correct scientific notation as $n \times 10^p$ is the defined form where $1 \le n < 10$. This example has

$n = 48$ which is greater than ten.

47. a) Saturn is about ten times as far from the sun as Earth.

b) Pluto is about forty times as far from the sun as the earth is.

c) Mercury is about 0.4 times as far from the sun as the earth.

49. a) A guess might be done by considering there are approximately 100,000 seconds in a day, so 1,000,000

seconds would be about 10 days. So a million seconds is about a week and a half.

b) $24 \text{ hours} / \text{day} \times 60 \text{ minutes} / \text{hour} \times 60 \text{ sec} / \text{minute} \approx 86,400 \text{ sec} / \text{day}$

$1,000,000 \text{ sec} \div 86400 \text{ sec} / \text{day} \approx 11.574 \text{ days}$

51. a) $\underbrace{1.59 + 0.30}_{\$2} + \underbrace{3.10 + 1.15}_{\$4} + \underbrace{0.72 + 2.00}_{\$3} + \underbrace{1.59}_{\$2} + \underbrace{0.89 + 2.29}_{\$3} \approx \$14$

b) $\underbrace{3.71 + 2.62}_{\$6} + \underbrace{0.51 + 0.30}_{\$1} + \underbrace{26.95 + 9.98}_{\$40} + \underbrace{4.25}_{\$4} \approx \$51$

53. The exact pay could be calculated mentally by $9.25 \times 20 = 9.25 \times 10 \times 2$
$$= 92.50 \times 2$$
$$= \$185$$

55. a) TRUNC(46.81792) = 46

b) TRUNC(-278.987) = -278

c) TRUNC(4.325) = 4

d) Truncating produces a different result from rounding when the tenths place has a 5, 6, 7, 8, or 9.

57. For student

Section 7.2

LE 1   Opener

You can change 0.3 to $\dfrac{3}{10}$ and 0.4 to $\dfrac{4}{10}$ and add using the understood rules for addition of fractions. A

second option is to create fraction squares with 0.3 and 0.4 to give that representation of the sum.

LE 2   Connection

a)

0.3 + 0.4

b) You add 3 units and 4 units to obtain 7 units.

c) The units are different, tenths rather than ones.

LE 3   Connection

A take – away approach of $0.3 - 0.1$ could be shown with two decimal squares. The first square is 0.3 and

the second square shows the difference after 0.1 has been taken away.

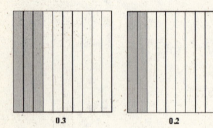

0.3   0.2

As an equation this is $0.3 - 0.1 = 0.2$

LE 4   Connection

Compare groups and measures is illustrated.

LE 5   Opener

The child can multiply $3 \times \dfrac{12}{100} = \dfrac{36}{100}$ or decimal squares can be used to show repeated addition,

$3 \times 0.12 = 0.12 + 0.12 + 0.12 = 0.36$

LE 6   Connection

a) 0.6×0.4 using decimal square

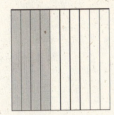 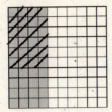

b) $0.6 \times 0.4 = \dfrac{6}{10} \times \dfrac{4}{10} = \dfrac{24}{100}$

c) Six tenths multiplied by four tenths is twenty – four hundredths

LE 7   Skill

a) The total costs of 12 shirts that cost $7.95 each is (a) $95.40

b) Estimation would round $7.95 to $8 and the product of eight and twelve is ninety – six

LE 8   Reasoning

a)

	Numbers			No. of Decimal Places	
Factor 1	Factor 2	Product	Factor 1	Factor 2	Product
3	0.12	0.36	0	2	2
0.6	0.4	0.24	1	1	2
12	7.95	95.4	0	2	2

b) The number of decimal places in the product is the sum of the numbers of decimal places in the two factors.

LE 9   Connection

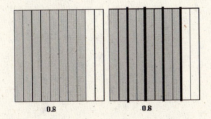

$0.8 \div 4 = 0.2$

LE 10   Reasoning

a)

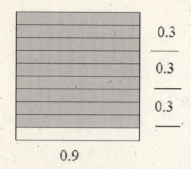

0.3

———

0.3

———

0.3

———

0.9

b) $0.9 \div 0.3 = 3$

$9 \div 3 = 3$

$90 \div 30 = 3$

$900 \div 300 = 3$

c) The quotient remains the same

d) $0.9 \div 0.3 = \dfrac{0.9}{0.3} = \dfrac{0.9}{0.3} \times \dfrac{10}{10} = \dfrac{9}{3} = 9 \div 3$

e) $a \div b = \dfrac{a}{b} = \dfrac{a}{b} \times \dfrac{c}{c} = \dfrac{ac}{bc} = ac \div bc$

LE 11   Concept

a) The divisor and dividend are multiplied by ten.

b) $0.08 \div 0.4 = \dfrac{0.08}{0.4}$ to convert this $\dfrac{0.08}{0.4} \times \dfrac{10}{10} = \dfrac{0.8}{4}$

c) $0.08 \div 0.4 = \dfrac{0.08}{0.4} = 0.2$

LE 12   Skill

This is approximately 6000 divided by one-third, or 18000

LE 13   Connection

a) $\dfrac{3}{11}$ means three divided by eleven

b)
$$11\overline{)3.00000} \quad \text{0.27272}$$

This is a repeating decimal due to the 27 repeating pattern

$$
\begin{array}{r}
0.27272 \\
11\overline{)3.00000} \\
\underline{22} \\
80 \\
\underline{77} \\
30 \\
\underline{22} \\
8
\end{array}
$$

LE 14   Connection

This illustrates partition, measures

LE 15   Reasoning

a)

$$
\begin{array}{ccccc}
0.6 & 0.8 & 0.7 & 0.9 & 0.5 \\
\underline{+0.3} & \underline{+0.9} & \underline{+0.6} & \underline{+0.2} & \underline{+0.8} \\
0.9 & 0.17 & 0.13 & 0.11 & 0.13
\end{array}
$$

b) The sum from the tenths column is all placed to the right of the decimal point.

c) Estimation could be used if for example we look at the second example $\begin{array}{r}0.8\\ \underline{+0.9}\\ 0.17\end{array}$, $0.8 > 0.17$ so if we add to

0.8 the sum must be greater than 0.17.

LE 16   Reasoning

a)

$$
\begin{array}{ccccc}
0.6 & 0.3 & 0.7 & 0.8 & 0.6 \\
\underline{+0.9} & \underline{+0.2} & \underline{+0.3} & \underline{+0.7} & \underline{+0.4} \\
5.4 & 0.6 & 2.1 & 5.6 & 2.4
\end{array}
$$

b) The product is given the same number of decimal places as each of the factors

c) Estimation could be used if for example we look at the first example $0.6 \times 0.9 \approx 0.6 \times 1 = 0.6$, so the

product of 5.4 is grossly incorrect

LE 17   Reasoning

a)

$$0.3\overline{)3.21}^{\;17} \quad 5\overline{)15.25}^{\;3.5} \quad 0.38\overline{)5.672}^{\;.79} \quad 4\overline{)36.16}^{\;9.4}$$

b) Zeros are omitted from the quotients

c) $0.3 \times 17 \approx \dfrac{1}{3}$ of $17 \neq 3.21$

LE 18   Skill

a) Susan's monthly earnings for February

b) It's the sum of the numbers from cells B1 to B12, inclusive

c) (B13)/12

LE 19   Summary

When adding decimals line up the decimal points and add (ignoring the decimal points) and insert the decimal point in the sum directly below those in the addends. When subtracting decimals line up the decimal points, subtract the numbers (ignoring decimal points) and insert the decimal point in the difference directly below those in the other two numbers. When multiplying terminating decimals, multiply the numbers. The number of decimal places in the product is the sum of the number of decimal places in the two factors. When dividing decimals, divide ignoring the decimal point, and place the decimal point in the quotient directly over the decimal point in the dividend.

7.2 Homework Exercises

1.

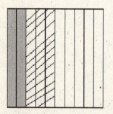

$0.2 + 0.3 = 0.5$

3.   a) When you add $6 + 7$ and separate the 13 into 3 and 10, this 10 represents 10 hundredths.

b) The 1 that is regrouped represents 1 tenth

c) The parts in (a) and (b) are the same

5.   Change the orientation of the problem from horizontal to vertical and compare results.

$$0.4 \text{ Line up the decimal point, don't forget to regroup.}$$
$$\underline{+0.8}$$
$$1.2$$

7.

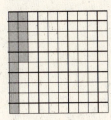

$0.23 - 0.08 = 0.15$

9.   Subtraction and take – away a measure/group

11.   You could present the examples b, a, then c. Choice b is first since there is no regrouping.

13.   a)

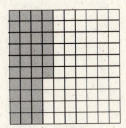

$2 \times 0.18 = 0.36$

b) $2 \times \dfrac{18}{100} = \dfrac{36}{100}$

c) Two times eighteen – hundredths equal thirty – six hundredths

15.   a)

The first square has 0.5 shaded the second square has 0.6 cross hatched, the intersection of shade and cross hatch is the product which is 0.3.

b) $0.5 \times 0.6 = \dfrac{5}{10} \times \dfrac{6}{10} = \dfrac{30}{100}$

c) Five tenths multiplied by six tenths equals thirty hundredths

17.   a) 1.2×0.4 is less than 1.2

b)

 +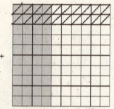

$(1 \times 0.4) + (0.2 \times 0.4) = 0.48$

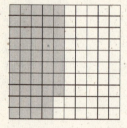

19.

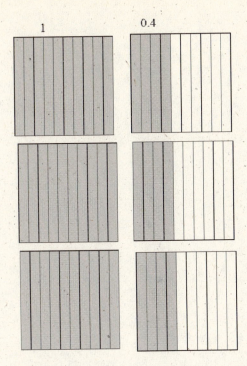

21.   a) With estimation, $34.56 \times 6.2 \approx 30 \times 6 = 180$, place the decimal point after the 4, 214.272 as this number is

closest to 180.

b) There are three digits behind the decimal in the factors of the products. The must be three digits behind

the decimal in the product, 214.272.

23.   a) A 60 – kg runner burns $0.12 \times 60 \times 10 = 72$ calories in a ten minute run

b) The illustrates equal measures

25. $50 \times 4.44 = \dfrac{50}{100} \times 4.44(100) = 0.5 \times 444$

27.   Shade in 0.6 then separate the decimal square into three equal parts.

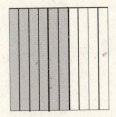

$0.6 \div 3 = 0.2$

29.   Shade 0.2 then determine how many groups of 0.04 can be made from 0.2.

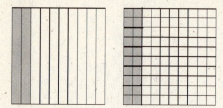

There are five groups, $0.2 \div 0.04 = 5$

31. $6.4 \div 0.32 = 6\dfrac{4}{10} \div \dfrac{32}{100}$

$= \dfrac{64}{10} \times \dfrac{100}{32}$

$= \dfrac{640}{32}$

$= 640 \div 32$

33.   Parts (a), (c), and (d) are equal.

$8 \div 0.23 = \dfrac{8}{0.23} \times \dfrac{10}{10}$ $\qquad\qquad$ $0.8 \div 0.023 = \dfrac{0.8}{0.023} \times \dfrac{10}{10}$

$\qquad\quad = \dfrac{80}{2.3}$ $\qquad\qquad\qquad\qquad\qquad = \dfrac{8}{0.23}$

$\qquad\quad = 80 \div 2.3$ $\qquad\qquad\qquad\qquad\quad = 8 \div 0.23$

35.   a) The child may have placed the remainder of one into the quotient.

b) The child might not understand that the 16 needs to have a decimal point so that the division can be

continued until it is determined that the quotient is terminating, non – terminating, or repeating.

$3\overline{)16.00000}$

37.   With estimation $44.237 \div 3.1 \approx 45 \div 3 = 15$ so the decimal point should be placed between the four and the

two, 14.27

39.   This is approximately 200 divided by one-fourth or 900

41.   You could present the examples a, c, then b. Choice a has the least amount of regrouping.

43.   Division and equal measures are illustrated

45.   Multiplication, equal groups/measures, addition, and combine groups measures are illustrated

47.   a) $4 - 0.03 = 3.97$

        b) $0.20 \times .3 \times 0.5 = \dfrac{2}{10} \times \dfrac{3}{10} \times \dfrac{5}{10}$

        $ = \dfrac{30}{1000}$

        $ = 0.030$

        c) $0.5 \div 0.02 = \dfrac{5}{10} \times \dfrac{100}{2}$

        $ = 5 \times 5$

        $ = 25$

49.   a) 2.5 pounds of salmon would cost $2.5 \times 4.99 = \$12.475$

        $ \approx \$12.48$

        b) 0.82 pounds of salmon would cost $0.82 \times 4.99 = \$4.0918$

        $ \approx \$4.09$

        c) It might the case that since 0.82 is less than 1 adults do not think about multiplication

51.   a) If $0 < x < 1 < y$ then $xy < y$ (Try $x = \dfrac{1}{2}, y = 2$)

        b)) If $0 < x < 1 < y$ then $\dfrac{y}{x} > y$ (Try $x = \dfrac{1}{2}, y = 2$)

        c) If $0 < x < 1 < y$ then $x^2 < x$ (Try $x = \dfrac{1}{2}$)

53.   a)
42	8.1	63	4.2
-3.71	-3.71	-5.29	-3.17
39.71	4.41	58.29	1.17

        b) The error occurs because of the lack of understanding of subtraction on the right hand side of the

        decimal point. There is no subtraction in the hundredths place in any of these examples and subtraction

        only occurs in the tenths place when there is digit in the tenths place for the minuend, otherwise the

        subtrahend decimal values are carried directly into the quotient.

55.   a)   16.2   14.1   12.3   8.2
            -3.7 -2.5 -6.7 -4.8
            13.5   12.4   6.4   4.6

b) The error is the subtraction on the decimal side, the smaller digit is subtracted from the larger digit without regard for regrouping and appropriate decimal subtraction

57.   United Kingdom: $\left(2.3\times10^{8}\right)\div\left(6.1\times10^{7}\right)=3.77$ tons

United States: $\left(2.3\times10^{9}\right)\div\left(3.02\times10^{8}\right)=7.61$ tons

France: $\left(2.6\times10^{8}\right)\div\left(6.3\times10^{7}\right)=4.13$ tons

59.   a) $1\times2+0.25=\left(1.5\right)^{2}$
           $2\times3+0.25=\left(2.5\right)^{2}$

b) $1\times2+0.25=\left(1.5\right)^{2}$
    $2\times3+0.25=\left(2.5\right)^{2}$
    $3\times4+0.25=\left(3.5\right)^{2}$

c) Yes, the equation is true.

61. Let x represent the miles per gallon for the family car, then $x-6$ is the miles per gallon for the SUV. The family car will need $\dfrac{10000}{x}$ gallons of fuel at a cost of $\$3.50\left(\dfrac{10000}{x}\right)$. The SUV will need $\dfrac{10000}{x-6}$ gallons of fuel at a cost of $\$3.50\left(\dfrac{10000}{x-6}\right)$.

a) The difference in fuel use for the SUV and the family car is

$$\frac{10000}{x-6}-\frac{10000}{x}=\frac{10000x-10000(x-6)}{x(x-6)}$$

$$=\frac{10000x-10000x+60000}{x(x-6)}$$

$$=\frac{60000}{x(x-6)}$$

b) The difference in fuel bill for the SUV and the family car is

$$\$3.50\left(\frac{10000}{x-6}\right)-\$3.50\left(\frac{10000}{x}\right)=\$3.50\left(\frac{10000x-10000(x-6)}{x(x-6)}\right)$$

$$=\$3.50\left(\frac{10000x-10000x+60000}{x(x-6)}\right)$$

$$=\$3.50\left(\frac{60000}{x(x-6)}\right)$$

63. You could multiply 11.2 by 2

65. a) A plan could include:

$ab.cd$ is the actual price and $cd.ab$ is the charged price. We're told that

$\$20.00 - cd.ab = $ correct change $+ \$11.88$

b) Possible solutions are \$12, \$13.01, \$14.02, \$15.03, \$16.04, \$17.05, \$18.06, \$19.07

67. a) $\dfrac{1}{3}=0.\overline{3}$

b) $\dfrac{1}{3}$ in base three is 0.1_{three}

c) $\dfrac{1}{3}$ in base six is 0.2_{six}

d) $\dfrac{1}{3}$ terminates in base nine, 0.3_{nine}

69.   a) D4 represents the amount of rent

b) The formula for F4 is F4 = F3 – D4

c)

checkbook

Check #	Date	For	Withd.	Deposit	Balance
Starting					$3,426.10
123	12-Mar	rent	$950		$2,476.10
	14-Mar			$875.40	$3,351.50
124	16-Mar	electric	$85.11		$3,266.39
125	17-Mar	shoes	$62.25		$3,204.14

71.   a) Watching TV $0.3 \times 2 \times \$0.08 = \$0.048$

b) Laundry $0.2 \times \$0.08 + 3 \times \$0.08 = \$0.256$

c) Refrigerator $120 \times \$0.08 = \$9.60$

d) Water Heater $350 \times \$0.08 = \$28$

e) Air conditioner $1.5 \times 8 \times \$0.08 = \$0.96$

Section 7.3

LE 1   Opener

a) Count the number of male and female students in your class today

b) $\dfrac{\text{males}}{\text{females}} =$

c) $\dfrac{\text{females}}{\text{females} + \text{males}} =$

LE 2   Connection

a)

A	B	B	B

b) Section A is $\frac{1}{4}$ of the whole sandwich

c) Section B is three times as long as section A

d) Section A is $\frac{1}{3}$ times as long as section B

LE 3   a) There are 750 more students than professors

b) There are 31 times as many students as professors

c) There are 46 times as many students as professors at Brain State.

LE 4   Skill

a) The ratio of miles to hours is $\frac{33}{6}$ and this is a measure of how far Kristin can ride per hour.

b) $\frac{11}{2} = 5.5$mph

c) Bill's rate $\frac{42}{8} = 5.25$mph is slower than Kristin's 5.5 mph

LE 5   Skill

Rate Table

Time (hours)	1	2	3	4
Distance Traveled (miles)	$5.5 \times 1 = 5.5$	$5.5 \times 2 = 11$	$5.5 \times 3 = 16.5$	$5.5 \times 4 = 22$

LE 6   Concept

       a) Class A $10:15$, Class B $20:30$

       b) Class A $2:3$, Class B $2:3$

       c) $\dfrac{10}{15} \times \dfrac{2}{2} = \dfrac{20}{30}$

       d) Class A $\dfrac{10}{15}$ Class B $\dfrac{20}{30}$

       $10 \times 30 = 15 \times 20$

       $300 = 300$

LE 7   Reasoning

       a) $\dfrac{a}{b} \bullet \dfrac{d}{d} = \dfrac{ad}{bd}$ $\dfrac{c}{d} \bullet \dfrac{b}{b} = \dfrac{cb}{db}$ $\dfrac{ad}{bd} = \dfrac{cb}{bd}$ and $bd = db$ by commutativity

       b) If $\dfrac{ad}{bd} = \dfrac{cb}{bd}$ then $ad = bc$

LE 8   Connection

       a) $\dfrac{10\,\text{ft}^3}{30\,\text{lb}} = \dfrac{40\,\text{ft}^3}{x\,\text{lb}}$

       b) 40 ft^3 of bricks would weigh 120 lbs.

       c) $\dfrac{10\,\text{ft}^3}{30\,\text{lb}} = \dfrac{40\,\text{ft}^3}{x\,\text{lb}}$

          $10x = 1200$

            $x = 120$

       d) Per ft^3 the bricks weigh: $\dfrac{40\,\text{ft}^3}{120\,\text{lb}} = \dfrac{1\,\text{ft}^3}{3\,\text{lb}}$ so a cubic foot of bricks weighs three pounds.

LE 9   Skill

a) You will need approximately twice as much seaweed as 7 people is approximately 8 people and the

original recipe is for 4 people.

b) $\dfrac{4\,\text{people}}{18\,\text{ounces}} = \dfrac{7\,\text{people}}{x\,\text{ounces}}$

$\qquad\qquad 4x = 126$

$\qquad\qquad x = 31\dfrac{1}{2}\,\text{ounces}$

c) The unit rate is in units of seaweed per person: $\dfrac{18\,\text{ounces}}{4\,\text{people}} = \dfrac{9\,\text{ounces}}{2\,\text{people}}$

$\qquad\qquad\qquad\qquad\qquad\qquad\qquad\qquad\qquad = 4\dfrac{1}{2}\,\dfrac{\text{ounces}}{\text{person}}$

Since there are seven people you will need $7 \times 4\dfrac{1}{2} = 31\dfrac{1}{2}$ ounces of seaweed

d)

Person	Ounces
1	$1 \times 4\dfrac{1}{2} = 4\dfrac{1}{2}$ ounces of seaweed
2	$2 \times 4\dfrac{1}{2} = 9$ ounces of seaweed
3	$3 \times 4\dfrac{1}{2} = 13\dfrac{1}{2}$ ounces of seaweed
4	$4 \times 4\dfrac{1}{2} = 18$ ounces of seaweed
5	$5 \times 4\dfrac{1}{2} = 22\dfrac{1}{2}$ ounces of seaweed
6	$6 \times 4\dfrac{1}{2} = 27$ ounces of seaweed
7	$7 \times 4\dfrac{1}{2} = 31\dfrac{1}{2}$ ounces of seaweed

e) There is no data to suggest a method for subsequent dinner invitations

LE 10   Skill

$\dfrac{3\,\text{male}}{2\,\text{female}}$, with 1100 students ask yourself how many groups of five are there, the sum of three and two,

$5\overline{)1100}\,^{220}$, So $\begin{array}{l} 3 \times 220 = 660\,\text{males} \\ 2 \times 220 = 440\,\text{females} \end{array}$ and the sum of 660 and 440 is 1100.

LE 11   Reasoning

a) Count your handful of white beans

b) Now count the same number of black beans and replace in bag from which you withdrew the white beans

c) When you remove a handful of beans from the mixed up bag consider a proportion to determine the number of beans in the bag

LE 12   Summary

Proportions are very useful to solve a problem with one unknown.

7.3 Homework Exercise

1.   a) The ratio of girls to boys is 17 to 12

b) The ratio of boys to all children is 12 to 29

3.   If the ratio of boys to girls is 4 to 5, then the student must understand that for each group of 9 children there are 4 boys.

5.   a)

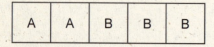

b) $\frac{2}{5}$ of the whole sandwich is section A

c) Section B is $1\frac{1}{2}$ times as long as section A

d) Section A is $\frac{2}{3}$ times as long as section B

7.   If one fifth of a college class is men then the ratio of men to women is 1 to 4.

9.   a) An additive comparison of their ages: The mother is 20 years older than the daughter

      b) A multiplicative comparison of their ages: The mother is twice as old as the daughter

11.   a) Round to the nearest hundredth

Student Teacher Ratio
18.01 : 1
13.83 : 1
16.10 : 1
15.24 : 1
14.51 : 1

      b) Atlanta has the lowest student teacher ratio

      c) San Francisco has the highest student teacher ratio

      d) Your response is a personal preference

13. $\frac{7}{4} \overset{?}{=} \frac{6}{3}$ no, as $7 \times 3 \neq 6 \times 4$

15.   a) If the number of members in each set is doubled the ratio will not change. Think about doubling a

      fraction then reducing it to lowest terms.

      b) If each set is increased by 10 then the ratio will change. 10 is a common addend, not a common factor.

17.   Looking at the proportions as fractions: $\frac{10}{3} = 3\frac{1}{3}$ people per platter , $\frac{14}{4} = 3\frac{1}{2}$ people per platter so the 14 to

      4 allots more food per person

19.   a) The population density is:

      New York: 19.4 million to 47,214 square miles is 410.90 people per square mile

      California: 37.4 million to 155,959 square miles is 239.81 people per square mile

      b) New York has a higher population density

      c) Visit the web and do a local search

21. a) The unit rate for each job is

$$\frac{\$500}{40\,\text{hours}} = \$12.50\,\text{per hour}$$

$$\frac{\$450}{36\,\text{hours}} = \$12.50\,\text{per hour}$$

b) The pay is identical

c)

Number of Hours	1	10	20	30	40
Job 1 Pays ($)	$12.50×1 = $12.50	$12.50×10 = $125.00	$12.50×20 = $250.00	$375	$500
Job 2 Pays ($)	$12.50×1 = $12.50	$12.50×10 = $125.00	$12.50×20 = $250.00	$375	$500

23. $2.29 / 32 oz, You pay $2.29 for 32 ounces of orange juice

$$\frac{\$2.29}{32\,\text{ounces}} \approx \frac{\$0.072}{\text{ounces}} \quad \text{you pay approximately seven cents per ounce of orange juice}$$

25. a) The unit price of this peanut butter is $\dfrac{\$2.79}{18\,\text{ounce}} = \dfrac{\$0.155}{\text{ounce}}$

b) The unit price of this peanut butter is $\dfrac{\$3.79}{32\,\text{ounce}} \approx \dfrac{\$0.12}{\text{ounce}}$

c) The first jar of 18 ounces has the lowest unit price

27. a) 3 red to 4 yellow with the addition of two more groups becomes 9 red to 12 yellow

b) $\dfrac{3}{4} = \dfrac{9}{12}$

29. Let c represent the number of comic books a child would receive for trading in 35, $\dfrac{4}{5} = \dfrac{c}{35}$.

$$140 = 5c$$
$$28 = c$$

A child could use the fundamental law of fractions or cross multiply to solve.

31. The comic book problem is a problem of proportions not of subtraction.

33.   a) $\dfrac{5}{7} = \dfrac{R}{14}$

         $R = 10$

    b) $\dfrac{3}{R} = \dfrac{6}{10}$

         $R = 5$

    c) $\dfrac{R}{45} = \dfrac{2}{6}$

         $R = 18$

35.   a) $\dfrac{Q}{5} = \dfrac{7}{3}$ $\dfrac{Q}{5} = \dfrac{7}{3}$

         $Q \approx 11$ $3Q = 35$

                               $Q = 11.\overline{6}$

    b) $\dfrac{Q}{12} = \dfrac{4}{Q}$ $\dfrac{Q}{12} = \dfrac{4}{Q}$

         $Q \approx 7$ $Q^2 = 48$

                               $Q = \pm\sqrt{48}$

                               $Q = \pm 4\sqrt{3}$

    c) $\dfrac{7}{Q} = \dfrac{5}{9}$ $\dfrac{7}{Q} = \dfrac{5}{9}$

         $Q \approx 12$ $63 = 5Q$

                           $12.6 = Q$

37.   If $\dfrac{5}{7} = \dfrac{N}{2}$ then $\dfrac{7}{5} = \dfrac{2}{N}$, $\dfrac{7}{2} = \dfrac{5}{N}$, $\dfrac{2}{7} = \dfrac{N}{5}$

39.　　a)　$\dfrac{x}{2} = \dfrac{3}{5}$

$$2\left(\dfrac{x}{2}\right) = 2\left(\dfrac{3}{5}\right)$$

$$x = \dfrac{6}{5}$$

　　　b)　$\dfrac{4}{y} = \dfrac{15}{2}$

$$\dfrac{y}{4} = \dfrac{2}{15}$$

$$4\left(\dfrac{y}{4}\right) = 4\left(\dfrac{2}{15}\right)$$

$$y = \dfrac{8}{15}$$

　　　c)　$\dfrac{3}{4} = \dfrac{10}{n}$

$$\dfrac{4}{3} = \dfrac{n}{10}$$

$$10\left(\dfrac{4}{3}\right) = 10\left(\dfrac{n}{10}\right)$$

$$\dfrac{40}{3} = n$$

41.	a) In four days they might drink about 1.5 quarts of orange juice and 2 quarts of skim milk, 4 days is about half of a week.

b) Orange Juice: $\dfrac{3 \text{ quarts}}{7 \text{ days}} = \dfrac{oj \text{ quarts}}{4 \text{ days}}$

$$12 = 7oj$$
$$1\dfrac{5}{7} = oj$$

Skim Milk: $\dfrac{2 \text{ quarts}}{7 \text{ days}} = \dfrac{sm \text{ quarts}}{4 \text{ days}}$

$$8 = 7sm$$
$$1\dfrac{1}{7} = sm$$

c) A unit rate for orange juice: $\dfrac{3}{7}$ quart per day, so in 4 days $4\left(\dfrac{3}{7}\right) = 1\dfrac{5}{7}$ quarts in 4 days

A unit rate for skim milk: $\dfrac{2}{7}$ quart per day, so in 4 days $4\left(\dfrac{2}{7}\right) = 1\dfrac{1}{7}$ quarts in 4 days

d) A multiplier scale for orange juice is $\dfrac{3}{7}$ quart per day then multiply by 4

A multiplier scale for skim milk is $\dfrac{2}{7}$ quart per day then multiply by 4

43.	a) Let x represent the price per pound $\dfrac{\$9}{\frac{3}{4}\text{lb}} = \dfrac{\$x}{1\text{lb}}$

$$9 = \dfrac{3}{4}x$$
$$\$12 = x$$

b)

$3	$3	$3		So the whole would be	$3	$3	$3	$3

45.	The woman is approximately five foot ten inches tall.

47.	Let p represent the weight of the eight foot length: $\dfrac{12'}{140\,\text{lbs}} = \dfrac{8'}{p}$

$$12p = 1120$$
$$p = 93\dfrac{1}{3}$$

49. The proportion of Francisco to Melissa is $\dfrac{3000}{5500}$ So Francisco's share of the business of $10,000 is

$\dfrac{3000}{8500} \times 10,000 \approx \$3,529.41$. Notice the denominator is $8,500, the sum of the initial investments.

51. 2,000,000 cars with 1.2 people per auto is 2,400,000 people. If the number of persons per auto increased to 1.5, then $2,400,000 \div 1.5 = 1,600,000$ autos would be needed.

53. Let f represent the number of fish in the lake: $\dfrac{10}{f} = \dfrac{6}{20}$ so there are approximately 34 fish in this lake.

$$200 = 6f$$
$$33\dfrac{1}{3} = f$$

55. Let t represent tea: $\dfrac{8}{5.30} = \dfrac{0.5}{t}$ so the amount of needed tea would cost you about $0.33.

$$1.895 = 8t$$
$$\$0.33125 = t$$

Let s represent sugar: $\dfrac{80}{4.99} = \dfrac{27.7}{s}$ so the amount of sugar would cost you about $1.73.

$$138.223 = 80s$$
$$\$1.727788 = s$$

Your total cost to create this product would be about $2.06.

57. "Fuel efficient" refers to the vehicle as a mode of transportation for passengers. Since 6 is $\dfrac{1}{4}$ of 24, the bus would be more efficient than the vehicle if it were carrying at least 4 times as many passengers. The same thinking applies to the train, since 2 is $\dfrac{1}{3}$ of 6, then if the train were carrying at least 3 times as many passengers as the bus it would be the most efficient. Comparing the train to the auto, since 2 is $\dfrac{1}{12}$ of 24, then the train would need to carry at least 12 times as many passengers as the auto.

59. On H gallons this car can travel $\dfrac{M(miles)H(gallons)}{G(gallons)}$ which results in final units of miles.

61. The function is cost based on the quantity of salmon purchased. Let C represent cost and S represent quantity of salmon purchased then $C = \$5S$.

63. Unit prices will vary depending on local supermarket and quantities of items available.

Section 7.4

LE 1   Reasoning

a) On the basis of the preceding examples, a shortcut for changing a percent to a decimal involves dropping the percent sign and moving the decimal point two places to the left to compensate.

b) To divide 100, move the decimal point 2 places to the left.

LE 2   Skill

a) The result of Example 1 suggests that a shortcut for changing a decimal to a percent involves moving the decimal point two places to the right.

b) $\frac{1}{8} = 0.125$, Moving the decimal point two places to the right for 0.125 yields 12.5%

LE 3   Connection

a) Forty – eight students out of eighty students play soccer

b) $\frac{48}{80} = 0.6$ or 60% of the students play soccer

c) The fraction of students who do not play soccer is $\frac{32}{80}$

d) The ratio of students who play soccer to those who do not is forty – eight to thirty – two.

LE 4   Opener

a) $0.28 \times 24000 = \$6720$

b) 14 out of 20 is 70% and 18 out of 25 is 72%, so the 18 out of 25 is a better percentage.

c) The regular price p is: $p - 0.80p = \$13$
$$0.2p = 13$$
$$p = \$65$$

LE 5   Skill

a) $\dfrac{18}{75} = \dfrac{6}{25}$ and $\dfrac{6}{25} = 0.24$ or 24%

b) 80% of 12 is $0.80 \times 12 = 9.6$

c) 46 is 40% of 60, $46 = 0.40x$
$$115 = x$$

LE 6   Reasoning

$Part = P \bullet whole$ where P is the percent.

LE 7   Skill

a) $\dfrac{18}{75} = \dfrac{n}{100}$ 24%
$$75n = 1800$$
$$n = 24$$

b) $\dfrac{46}{n} = \dfrac{40}{100}$ The number is 115.
$$40n = 4600$$
$$n = 115$$

c) $\dfrac{n}{12} = \dfrac{80}{100}$ 24 is 30% of 80.
$$100n = 960$$
$$n = 9.6$$

LE 8   Reasoning

a) In one year you would earn $0.03(\$2000) = \$60$.

b) In three years you would earn $0.03(\$2000)(3) = \$180$, or three times as much is (a).

c) In half a year you would earn half of the value for one year, part (a), $30.

d) $I = \Pr t$

LE 9   Skill

In the account you have the original $1000 plus the interest,

$1000 + 0.04(\$1000)(2) = \$1080$

LE 10   Skill

a) Interest is $0.04(\$1000)(1) = \$40$ and the balance is $\$1000 + 0.04(\$1000)(1) = \$1040$

b) After two years the account balance would be $\$1040 + 0.04(\$1040)(1) = \$1081.60$

c) In two years you have $\$1081.60 - \$1080 = \$1.60$ more

LE 11   Summary

A shortcut for changing a percent to a decimal involves dropping the percent sign and moving the decimal point two places to the left to compensate. A shortcut for changing a decimal to a percent involves moving the decimal point two places to the right.

7.4 Homework Exercises

1.   For student

3.   a) 150% means 150 per 100

     b)

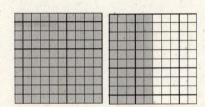

     c) 150% means 30 per 20

     d) 150% means 3 per 2

5.   a) 34% is $\dfrac{34}{100} = 0.34$

b) 180% is $\dfrac{180}{100} = 1.80$

c) 0.06% is $\dfrac{\frac{6}{100}}{100} = \dfrac{6}{10000} = 0.0006$

7.   A percent that is greater than 100% represents a number greater than one.

9.   a) 0.23 is 23%

b) 0.00041 is 0.041%

c) 24 is 2400%

11.   a)

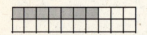

b) $\dfrac{1}{20} = 0.05$ or 5%

c) $\dfrac{7}{20} = 0.35$ or 35% you could also multiply the value of one square by 7 to obtain the same result

13.   a) $\dfrac{1}{25} = 0.04$ which is 4%

b) $\dfrac{3}{8} = 0.375$ which is 37.5%

c) $1\dfrac{3}{4} = 1.75$ which is 175%

15.   a) 132 out of 200 adults did not read a book

b) $\dfrac{132}{200} = 0.66$, 66% of the adults didn't read a book

c) The fraction of adults who did read a book is $\dfrac{68}{200}$

d) The ratio of adults who read to those who did not is 68 to 132

e) Adults who did not read books out number those who did read a book by 64

17.   495 out of 500 watched television in the past week

$\dfrac{495}{500}$ watched television in the past week

$\dfrac{495}{500} = 0.99$, 99% watched television in the past week

People that watched television this past week out numbered those who did not watch television by 490

19.   a) 30% of 140 is: $\dfrac{30}{100} = \dfrac{n}{140}$
                                    $100n = 4200$
                                    $n = 42$

        b) 25 is 40% of what: $\dfrac{25}{n} = \dfrac{40}{100}$
                                    $40n = 2500$
                                    $n = 62.5$

        c) 27 out of 40 is: $\dfrac{27}{40} = \dfrac{n}{100}$
                                    $40n = 2700$
                                    $n = 67.5$

21.   a) 120 is 60% of: $0.6n = 120$
$$n = 200$$

b) 36 out of 80 is: $36 \div 80 = 0.45$, 45%

c) 70% of 20 is: $0.70 \times 20 = 14$

23.   a) Consider that 12% is about 10% thus the commission is estimated with a decimal shift, so the

commission on $2400 is about $240

b) 12% of $2400 is: $0.12 \times 2400 = \$288$

c) $\dfrac{12}{100} = \dfrac{n}{2400}$
$$100n = 28800$$
$$n = 288$$

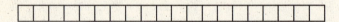

d) If she sells 3 times as many computers, her commission will be three times as much

25.   a) $82 \div 110 = 0.74\overline{54}$ or $74.\overline{54}\%$

b) $\dfrac{82}{110} = \dfrac{n}{100}$
$$110n = 8200$$
$$n = 74.\overline{54}\%$$

27.   a) Let s represent sales $0.08s = \$798.40$
$$s = \$9980$$

b) $\dfrac{8}{100} = \dfrac{798.40}{s}$
$$8s = 79840$$
$$s = \$9980$$

29.   To represent 100% consider that 40% is shown with 8 squares, so another 8 squares is 80% and four more

squares would be 100%. So the illustration has 20 squares.

31. a) First convert 296 million to billions, 0.296 for ease of calculation: U.S. $\dfrac{1895}{0.296} \approx \$6402.03$, Do the same

conversion for Canada: $\dfrac{96}{0.032} = \$3000$

b) U.S. Administrative cost $\dfrac{420}{1895} \approx 0.2216359$ or 22.16%

Canada Administrative cost $\dfrac{10.6}{96} \approx 0.11041667$ or 11.04%

c) 11% of \$1895 billion is: $0.11 \times 1895 = \$208.45$ billion. The difference between the current administrative

costs and the proposed costs is \$420 - \$208.45 = \$211.55 billion

d) According to the data there is a lower occurrence of infant mortality in Canada than the U.S. There are

fewer patients for U.S. doctors and Canadians have a longer life expectancy.

33. The company profit, P, is the difference between total sales, s, and expenses. $P = s - (0.4s + \$35000)$ If the

desire is to have a profit of \$40000 then s must be:

$40000 = s - (0.4s + \$35000)$ Sales must be at least \$125,000
$75000 = 0.6s$
$125000 = s$

Sales must be at least \$125,000.

35. $I = \text{Pr}t$

$I = \$5000(0.06)\left(\dfrac{1}{2}\right)$

$I = \$150$

37.   a) This company charges one twelfth of the annual rate per month, $\frac{1}{12}(0.12) = 0.01$ or 1%

b) Using a 360 day year the daily rate is $\frac{1}{360}(0.12) = 0.000\overline{3}$ or 0.033%

c) $660(0.01) = \$6.60$

39.   a) The interest is $I = \mathrm{P}r t$
$$I = \$8000(0.06)(1)$$
$$I = \$480$$

and the account balance would the sum of the interest and the principle, $8480

b) Let A represent the future amount $A = P + I$
$$A = P + \mathrm{P}r t$$
$$A = 8480 + 8480(0.06)(1)$$
$$A = \$8988.80$$

41.   a) The bank will pay half the amount of the annual interest or 2%

b) The interest is $I = \mathrm{P}r t$
$$I = \$6000(0.04)\left(\frac{1}{2}\right)$$
$$I = \$120$$

and the account balance would the sum of the interest and the principle, $6120.

c) Let A represent the future amount $A = P + I$
$$A = P + \mathrm{P}r t$$
$$A = 6120 + 6120(0.04)\left(\frac{1}{2}\right)$$
$$A = \$6242.40$$

43.   a) This is an example of an regressive tax structure.

b) This is an example of an intermediate tax structure.

c) 6% of $6000 is $360, 6% of $3000 is $180 now as a percentage of annual income $200 is what percent

of $45000, about 0.4% and $100 is what percent of $10000, about 1% so this is a regressive structure.

45.   a) After one year you have 104% of the original amount.

       b) You could multiply the principal by 1.04 to find the amount after one year.

       c) You would multiply by 1.04 to the second power.

       d) A quick way is $A = P(1+r)^t$
$$A = 1000(1+0.04)^8$$
$$A \approx \$1368.57$$

47.   a) The world population in 2009 $6.7 \times 1.011 \approx 6.7737$ billion

       b) The world population in 2020, 11 years later $6.7737 \times 1.011^{11} \approx 7.639918$ billion

       c) The world population in 2050, 41 years later $6.7737 \times 1.013^{41} \approx 11.503056$ billion

Section 7.5

LE 1   Skill

a) 50% is $\dfrac{50}{100} = \dfrac{1}{2}$, 25% is $\dfrac{25}{100} = \dfrac{1}{4}$, 10% is $\dfrac{10}{100} = \dfrac{1}{10}$, 1% is $\dfrac{1}{100}$

b) Computing 50% of a number is the same as dividing the number by two

c) Computing 25% of a number is the same as dividing the number by four

d) Computing 10% of a number is the same as dividing the number by ten

e) Finding 10% of a number is the same as dividing by ten, which means you could move the decimal point in the number one place to the left.

f) Finding 1% of a number is the same as dividing by 100, which means you could move the decimal point in the number 2 places to the left.

LE 2   Skill

a) 50% of 286 is the same as asking what is half of 286, 143

b) 25% of 4000 is the same as dividing 4000 by 4, 1000

c) 1% of 380 is the same as dividing 380 by 100, 3.8

LE 3   Skill

The discount is found by shifting the decimal point one place to the left $2.5 and the sale price is found by subtracting the discount from the original price, $25 - $2.5 is $22.50

LE 4   Skill

a) 70% of 300 could be mentally computed by first dividing 300 by 100 which is 3, then multiplying 70 by 3 which is 210

b) 6% of 900 could be mentally computed by first dividing 900 by 100 which is 9, then multiplying 9 by 6 which is 54.

LE 5   Reasoning

The first thing the student needs to realize is that their value is greater than the number they began with.

$$0.27(400) = (0.25 + 0.01 + 0.01)(400)$$
$$= (0.25)(400) + (0.01)(400) + (0.01)(400)$$
$$= 100 + 4 + 4$$
$$= 108$$

LE 6   Skill

Fraction	$\frac{1}{20}$		$\frac{1}{10}$	$\frac{1}{5}$	$\frac{1}{4}$	$\frac{1}{3}$	$\frac{1}{2}$	$\frac{2}{3}$	$\frac{3}{4}$	$\frac{1}{1}$
Percent	$\frac{1}{20} = 0.05$, 5%		10%	20%	25%	$33\frac{1}{3}\%$	50%	$66\frac{2}{3}$	75%	100%

LE 7   Skill

Rounding and breaking

22% of $648 is about 20% of 600 or $\frac{1}{5}$ or 600 or 120

Compatible numbers

22% is about 20% or $\frac{1}{5}$ and or $\frac{1}{5} \times 650 = 130$

LE 8   Skill

a) Her increase was $48300 - 43870 = \$4430$, then $\$4430 = P \bullet \$43870$ so her raise was about 10.1 %

$$\frac{4430}{43870} = P$$
$$0.10098 \approx P$$

b) $\frac{4430}{43870} = \frac{n}{100}$

   $43870n = 443000$

      $n \approx 10.098\%$

c) Nancy's increase was 12% so this increase is about 2% less

d) Margaret's increase was not larger than Nancy's on a percentage basis but if you were to compare dollar amounts then Margaret's was larger

e) How do you use the word "better?" If better is a reference to dollar amount or percentage, then that will decide your response to this question.

LE 9   Reasoning

Rico was probably comparing population increase, Harmony increased by 2,000 people and Pine Valley increased by 3,000 people.

Ananya was probably using percentages.

Harmony: $\frac{2,000}{10,000} = 0.20$ or 20% increase

Pine Valley: $\frac{3,000}{20,000} = 0.15$ or 15% increase

LE 10   Skill

a) 10% off would be $14.0 and three times that amount would be the 30% or $42.0 so the sale price is $98.

b) $98 is 70% of $140, 70% of $140 is $98.

LE 11   Summary

To mentally compute certain percentages it may be "easier" to compute with the fractional equivalent. Tips at a restaurant can be calculated by first finding 10% and moving from there to the desired percentage.

7.5 Homework Exercises

1.	a) 50% of 222 is the same as asking what is half of 222, 111

	b) 1% of 24 is the same as dividing 24 by 100, 0.24

	c) 10% of 470 is the same as shifting the decimal point in 470 one place to the left, 47.0

3.	a) 3% of 500 could be mentally calculated by first dividing 500 by 100, which is 5, then multiplying by 3 which is 15

	b) 80% of 400 could be mentally computed by first dividing 400 by 10 which is 40, then multiplying 40 by 8 which is 320

	c) 75% of 12 could be mentally computed by first dividing 12 by 4 which is 3, then multiplying 3 by 3 which is 9

5.	To estimate the number that voted out of 227,000,000 with a 41% approximation of participation consider that 227,000,000 is about 200,000,000 and 41% is about 40% so 40% of 200,000,000 is 80,000,000

7.	a) Using 1% of 400 is the same as dividing 400 by 100 or shifting the decimal two places to the left so the result is 4, then multiply this result by 8, 32

	b) Using 1% of 900 is the same as dividing 900 by 100 or shifting the decimal two places to the left so the result is 9, then multiplying this result by 7, 63

9.	a) 50% of 44 is 22 "Half of what is 22?"

	b) 25% of 320 is 80 Multiply 80 by 4 since that is the denominator of 25% in fractional form

	c) 20% of 550 is 110 Multiply 110 by 5 since that is the denominator of 20% in fractional form

11.	a) If that region represents 25% then double it to represent 50%

	b) 100% could be represented as

	c) 75% could be represented as

13. Since there are three juice ingredients and grape juice is the most abundant there is at least $33\frac{1}{3}$% grape

juice

15. a) To estimate the seating with rounding, 83,000 is about 80,000 and 64% of is the same as two decimal

shifts to the left for 80,000 then multiplying by 83, so there are about 66,400

b) To estimate with compatible numbers, round 83,000 to 84,000 since this value is divisible by three and

round 64% to $66\frac{2}{3}$% which is two – thirds as a decimal, so $\frac{2}{3}$ of 84,000 is 56,000

17. a) This result is unreasonable as 15% of 724 is less than 724

b) 20% of 58000 is about 12,000 so this result is reasonable

c) This result is unreasonable as 86% of 94 is less than 94

19. 48 is about 50, double 50 and you have 100, so double 34 you have 68, so Zine's percent was about 68%

21. 10% of 21.07 is about $2.10 and half of that is $1.05 so the 15% tip would be $3.15

7% is about half of 15% so double the tax amount which is about $1.50, or $3.00 tip

23. a) $16600 \times \dfrac{n}{100} = 16600 - 7000$

$$166n = 9600$$
$$n \approx 57.83\%$$

b) $\dfrac{16600 - 7000}{16600} = \dfrac{n}{100}$

$$\dfrac{9600}{16600} = \dfrac{n}{100}$$
$$16600n = 960000$$
$$n \approx 57.83\%$$

25. a) An additive comparison, the new salary is $3000 more

b) A multiplicative comparison, 10% of $30000 is $3000, the amount of the increase

27. a) At 20% off this $15.98 glove will sell for: 10% of the price is $1.60 with rounding so 20% is $3.20 and

the difference of $15.98 and $3.20 is $12.78, the sale price

b) If the sales tax is 4%, 1% of $12.78 is $0.13 with rounding and multiplying this by 4 yields $0.52 in tax

for a total cost of $13.30 if you use $12.78 \times 1.04 \approx $13.29

29. a) The new price is 1.06 times the old price

b) The new population is 1.12 times the old population

31.   a) The new price is 25% higher than the old price

        b) The old price is 20% lower than the new price

        c) The new price is 125% of the old price

        d) The old price is 80% of the new price

33.   Let's use $100 to judge equivalent statements

Part	Old Price	New Price
a)	100	120
b)	100	20
c)	100	120
d)	100	125
e)	100	120
f)	100	125

a), c) and e) are equivalent. d) and f) are equivalent

35.   a) Let n be the unknown, $30 = 0.7n$ so the child's work is incorrect

$$\frac{30}{0.7} = n$$
$$42.86 \approx n$$

b) The child's work is $0.3(30) + 30 = 39$ whereas the correct procedure is $30 = 0.7n$

$$\frac{30}{0.7} = n$$
$$42.86 \approx n$$

37.   The new price of the dress if w is the wholesale price is $1.25w = \$80$

$$w = \$64$$

so the dollar amount of the profit is the difference between 80 and 64, or \$16.

39.   Which shoes have a higher percent markdown? Write the percent markdown of these shoes in descending order.

41.   a) The prediction is that next year's prices will go up.

b) If b is 100 then a is 85. 15% more than 85 is 97.75 so b is not 15% greater than a

43.   a) 48% of 659 is close to 300

b) 48% of 659 is 316.32

c) 45% of 689 is close to 300

d) 45% of 689 is close to 310.05

45.   Let w represent the wholesale price so the price of the vehicle s is 1.3 times w, $s = 1.3w$, so now the sale price is 80% of s, $0.8s = 0.80(1.3w)$ so the sale price is 4% over the wholesale price

$$0.8s = 1.04w$$

47.   Two years ago the salary was s so after the first 20% increase, the salary was 1.2 times s. The salary after the second 20% increase is $1.2s + 0.20(1.2s) = 1.44s$ so solving for s when we know the current salary of

\$69840 yields: $\$69840 = 1.44s$

$$\$48500 = s$$

49.   This represents percent increase of $\dfrac{M - N}{N} \times 100\%$

51.   a) C3 represents the Rearguard dollar amount on 1-1-04

b) The D3 formula is C3 – B3. The E3 formula is (C3 – B3)/B3*100

c)

Fund	1/1/2008	1/1/2009	$ change	% change
Wings	$8,342.61	$8,680.10	$337.49	4.05
Rearguard	$12,471.82	$11,871.17	($600.65)	(4.82)

The parentheses in the spreadsheet represent negative amount

Section 7.6

LE 1   Opener

When we change a rational to a decimal the result is either a terminating or a repeating decimal, for example:

$\dfrac{1}{2}$ as a division problem $2\overline{)\,1}$ $\;\overset{0.5}{}$

$\dfrac{2}{3}$ as a division problem $3\overline{)\,2}$ $\;\overset{0.\overline{6}}{}$

LE 2   Concept

The decimal represents a rational fraction if it terminates or has a repeating pattern and is infinite

LE 3   Reasoning

a) $\dfrac{1}{9} = 0.\overline{1}$ $\dfrac{2}{9} = 0.\overline{2}$

b) Based on the "pattern" of part a, $\dfrac{7}{9} = 0.\overline{7}$

c) $\dfrac{1}{99} = 0.\overline{01}$ $\dfrac{2}{99} = 0.\overline{02}$

d) Based on the "pattern" of part c, $\dfrac{7}{99} = 0.\overline{07}$

e) Based on the "pattern" of part c, $\dfrac{13}{99} = 0.\overline{13}$ which is the result of a calculator

f) Based on these "patterns" a conjecture, $\dfrac{278}{999} = 0.\overline{278}$

LE 4   Skill

a) Using the general formula $0.\overline{8} = \dfrac{8}{10^1 - 1}$

$$= \dfrac{8}{9}$$

b) $0.\overline{37} = \dfrac{37}{10^2 - 1}$

$$= \dfrac{37}{99}$$

c) $0.\overline{02714} = \dfrac{2714}{10^5 - 1}$

$$= \dfrac{2714}{99999}$$

LE 5   Skill

a) $-0.82, -\dfrac{3}{4}, \dfrac{3}{4}, 0.82, 0.\overline{82}$

b) $-1, -\dfrac{2}{3}, -\dfrac{7}{12}, -0.5$

LE 6   Concept

$\sqrt{2}$ is a non-repeating decimal

LE 7   Concept

Another example of the square of a counting number having an even number of factors in its prime

factorization is $81 = 3 \bullet 3 \bullet 3 \bullet 3$

LE 8   Reasoning

a) Suppose $\sqrt{2} = \dfrac{p}{q}$ in which p and q are counting numbers. Square both sides, and you obtain $2 = \dfrac{p^2}{q^2}$

b) Solving the equation of part a), $2q^2 = p^2$

c) So $2q^2 = p^2$. Imagine prime factoring $2q^2$ and p^2. Since p^2 is the square of a counting number, it has

an even number of prime factors.

d) Each in this proof involves deductive reasoning

LE 9   Concept

a) Two more square roots that are irrational are $\sqrt{3}$ and $\sqrt{5}$

b) A square root that is a rational number is $\sqrt{4}$

LE 10   Reasoning

a) and b) Answers will vary depending on the size circles that students construct

c) The last column should show a pattern such that the ratio of $\frac{C}{d} \approx 3.14$

d) The generalization suggest that the ratio of circumference to diameter is pi.

LE 11   Skill

The approximate circumference for a wheel with a 30 inch diameter is about 90 inches. Using estimation

that pi is approximately 3 then the product of pi and the diameter will approximate the circumference.

LE 12   Concept

Two examples of real numbers are 3 and 8.765

LE 13   Concept

	Whole Number	Integer	Rational Number	Irrational Number	Real Number
2.6			√		√
3.00%	√	√	√		√
$\sqrt{2}$				√	√
−1		√	√		√
$-2\frac{1}{2}$			√		√

LE 14   Connection

a) The inequality for this number line is $n < 10$

b) $n \geq 5$ can be graphed as

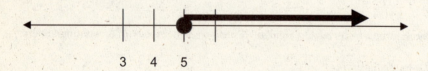

3   4   5

LE 15   Reasoning

$\frac{3}{7}$ may be easier than the decimal equivalent of $0.\overline{428571}$

LE 16   Reasoning

a) This addition may be "easier" using the fractions.

b) This addition may be "easier" using the decimals.

c) Fractions are "easier" when there is a common denominator that has factors other than 2 and 5. Decimals are "easier" for fractions that are easily converted to decimals.

LE 17   Connection

a) 10.3% is an approximation.

b) $10\frac{1}{3}\%$ is an exact response

c) The advantage of the exact answer is the precision it offers whereas the decimal approximation may be "easier" to work with.

LE 18   Connection

a) The decimal representation of $\sqrt{273}$ is a non – repeating decimal.

b) Since $\sqrt{273} \approx 16.5227$ you could tell the farmer to purchase 16.6 feet of fence, since 16.5 would be a little short.

LE 19   Concept

For example $0.2 + 0.3 = 0.3 + 0.2$

LE 20   Concept

This illustrates the additive inverse property

LE 21   Concept

a) The common factor is x, so $2x + x\sqrt{2} = x(2 + \sqrt{2})$

b) Distributive property of multiplication over addition.

LE 22   Reasoning

Since the whole numbers are a subset of the real numbers, a counter – example for whole numbers is also a counter – example for the real numbers.

LE 23   Concept

Zero and one are not elements of the set of irrationals therefore these cannot be the additive identity or multiplicative identity respectively

LE 24   Summary

   a) Rational numbers can be expressed as terminating decimals and repeating decimals. Non – terminating, non – repeating decimals are called irrational numbers. $\sqrt{2}$ is the most famous irrational number. The union of the set of rational numbers and the set of irrational numbers is the set of real numbers.

   b) Real numbers have the same number operations as rational numbers

Section 7.6

1.   a) $\dfrac{3}{5} = 0.6$

   b) $2\dfrac{2}{9} = 2.\overline{2}$

3.   a) $0.731 = \dfrac{731}{1000}$

   b) $-13.04 = -(13 + 0.04)$

      $= -(13 + \dfrac{4}{100})$

      $= -(13 + \dfrac{1}{25})$

      $= -13\dfrac{1}{25}$

5.   a) $0.\overline{4} = \dfrac{4}{10^1 - 1}$

      $= \dfrac{4}{9}$

   b) $0.\overline{32} = \dfrac{32}{10^2 - 1}$

      $= \dfrac{32}{99}$

   c) $0.\overline{267} = \dfrac{267}{10^3 - 1}$

      $= \dfrac{267}{999}$

      $= \dfrac{89}{333}$

7. a) $0.2\overline{1} = \dfrac{2\dfrac{1}{9}}{10}$

$= \dfrac{\dfrac{19}{9}}{10}$

$= \dfrac{19}{90}$

b) $0.3\overline{41} = \dfrac{3\dfrac{41}{99}}{10}$

$= \dfrac{\dfrac{338}{99}}{10}$

$= \dfrac{338}{990}$

9. a) If $N = 0.\overline{2}$ then $10N = 2.\overline{2}$

b) $10N = 2.\overline{2}$

$\underline{-\ N = 0.\overline{2}}$

$9N = 2$

c) $9N = 2$

$N = \dfrac{2}{9}$

d) $100N = 31.\overline{31}$

$\underline{-\ \ N = 0.\overline{31}}$

$99N = 31$

$N = \dfrac{31}{99}$

11. a) $-1, -0.4, -\dfrac{3}{10}, -\dfrac{1}{5}$

b) $-0.7, -\dfrac{2}{3}, 0.\overline{57}, 0.5\overline{7}, \dfrac{2}{3}$

13. If a square's area is 5 square units, then what is the side length of this square?

15. a) On a calculator the $\sqrt{5}$ is shown as 2.236067977, the number of decimal places shown is dependent on your calculator

b) This cannot be the exact value as it is a non − terminating, non − repeating decimal

17. The diameter of wheel with a circumference of 76 inches is found using $C = \pi d$, $76 = \pi d$

$$\frac{76}{\pi} = d$$

$$24.19 \approx d$$

The diameter is approximately 24.19 inches

19. a) $\sqrt{11}$ is irrational

b) $\frac{3}{7}$ is rational

c) π is irrational

d) $\sqrt{16}$ is rational

e) 50% is rational

21. a) 0.34938661… is irrational

b) $0.2\overline{6}$ is rational

c) 0.565665666…is irrational

d) 0.56 is rational

23. You might first ask the child to say in words what 0.2 is. Hopefully the response is two – tenths. Then write this out to show $\frac{2}{10}$ which is rational. For $\sqrt{49}$ ask the student what the square root of forty – nine is. If they respond seven, then show that seven is the same as $\frac{7}{1}$.

25. a) For a 54 foot skid mark on a wet highway the speed of the auto is approximately $s \approx \sqrt{12(54)}$

$$s \approx \sqrt{648}$$

$$s \approx 25.5 \text{ mph}$$

b) For a 54 foot skid mark on a dry highway the speed of the auto is approximately $s \approx \sqrt{24(54)}$

$$s \approx \sqrt{1296}$$

$$s \approx 36 \text{ mph}$$

c) The approximation in part a is irrational, the approximation in part b is rational

27. Terminating and repeating match with rational

Infinite non-repeating matches with irrational

29.   a) There are three whole numbers between -3 and 3, {0, 1, 2}

        b) There are five integers between -3 and 3, {-2, -1, 0, 1, 2}

        c) There are an infinite number of real numbers between -3 and 3

31.   0.3333333 could be $\dfrac{1}{3}$ or $\dfrac{3333333}{10000000}$

33.

	Whole Number	Integer	Rational Number	Irrational Number	Real Number
$\dfrac{1}{3}$			√		√
$\sqrt{13}$				√	√
−6		√	√		√
$\sqrt{9}$	√	√	√		√
−0.317			√		√

35.

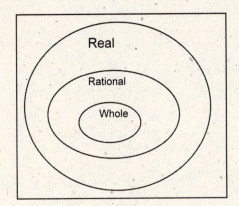

37.   a) $x \le 1$

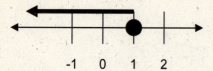

b) $x > -3$

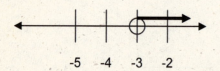

39. $\sqrt{2}$ is an example of a number that cannot be written as a rational or easily represented by a decimal.

41.   With a calculator you could convert each fraction to its respective decimal representation and then compare decimals.

43.   An addition problem that is "easier" with fractions than decimals could be $\frac{1}{3} + \frac{1}{6}$.

45.   a) $\frac{22}{7} \approx 3.142857143$ is a closer approximation.

b) On my calculator $\pi \approx 3.141592654$

47.   The symbol π is a non – terminating, non – repeating decimal

49.   a) Commutative property of addition

b) Associative property of multiplication

c) Distributive property of multiplication over addition

d) Additive Identity

51.   The additive inverse of -0.41798 is 0.41798

53.   A counterexample that real number subtraction is not commutative could be $7 - 2 \ne 2 - 7$

55.   Mentally you could multiply 6 and 8 then add half of 8, $6.5 \times 8 = \left(6 + \frac{1}{2}\right)8$

$$= 6(8) + \frac{1}{2}(8)$$
$$= 48 + 4$$
$$= 52$$

57. a) $x + y = 3$ is true for some pairs of real numbers

b) $xy = yx$ is true for all pairs of real numbers

c) $x + y = x + y + 3$ is true for no real numbers

d) $\sqrt{x^2 + y^2} = x + y$ is true for some pairs of real numbers

59. a) $\dfrac{4}{5}$ can be written with denominators of 10, 100, 1000 and so on

b) $\dfrac{4}{5}, \dfrac{3}{25}, \dfrac{3}{200}$ can be represented by terminating decimals

c) Fractions (in simplest form) that can be rewritten as terminating decimals have denominators that are divisible by no prime number other than 2 or 5

61. $\sqrt{M + N} = \sqrt{M} + \sqrt{N}$ works if $M = 0$, or $N = 0$ or both M and N are zero

63. a) $0.010110111... + 0.101001000... = 0.11111111...$ (notice where the zeros and ones occur in the two addends

b) Part (a) shows that the sum of two irrational numbers can be a rational number

65. Assume $\sqrt[3]{2}$ is rational. Then $\dfrac{p}{q} = \sqrt[3]{2}$ in which p and q are counting numbers. Then $p = q\sqrt[3]{2}$ and

$p^3 = 2q^3$. If p^3 is the cube of some counting number then the number of prime factors it has is a multiple of three. $2q^3$ has a number of prime factors equal to a multiple of three plus one. So p^3 and $2q^3$ are equal numbers that have a different number of prime factors, which is impossible. Therefore $\sqrt[3]{2}$ is irrational

67. a) $\dfrac{3}{16} = 0.1875$

b) $0.316 = \dfrac{316}{1000}$

$= \dfrac{79}{250}$

1.   0.0037 in expanded notation, $0.0037 = \left(3 \times \dfrac{1}{1000}\right) + \left(7 \times \dfrac{1}{10000}\right)$

3.   By extending a pattern $4^3 = 64$ Using the exponent addition rule $4^2 \bullet 4^0 = 4^{2+0}$

                                $4^2 = 16$ $= 4^2$

                                $4^1 = 4$ Therefore

                                $4^0 = 1$ $4^0 = 1$

5.   23.7 million in scientific notation is 2.37×10^7

7. $0.7 - 0.2 = 0.5$

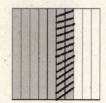

     c) 2 tenths times 4 tenths equals 8 hundredths.

9.   With estimation this problem is approximately 120 divided by 2 which is 60, so the decimal point in the

     original problem should be placed between the 1 and the 5, 61.5.

11.   a) $8 - 0.79 = 7.21$

     b) $0.3 \times 0.5 \times 0.8 = 0.3 \times 0.4$

                        $= 0.12$

     c) $0.076 \div 0.8 = \dfrac{76}{1000} \times \dfrac{10}{8}$

                      $= \dfrac{76}{800}$

                      $= \dfrac{19}{200}$

                      $= 0.095$

13.   The operation is division and the category is partition.

15.   Repeating the same error pattern, $6 \div 0.3 = 0.2$ $0.8 \div 0.4 = 0.02$ $1.2 \div 0.3 = 0.04$

17.   a) Six more students like recess than those who do not.

     b) Three times as many students like recess compared to those who do not

19. Let d represent the number of peanuts that Debbie ate, then Maria a quantity of $5d$ peanuts and together they ate 96 peanuts, $d + 5d = 96$ so Debbie at 16 peanuts and Maria ate 80 peanuts

$$6d = 96$$
$$d = 16$$

21. a) $$\frac{34}{100} = \frac{n}{40}$$
$$\frac{34 \cdot 40}{100} = n$$
$$13.6 = n$$

   b) $$\frac{3}{250} = \frac{n}{100}$$
$$\frac{300}{250} = n$$
$$1.2\% = n$$

   c) $$\frac{43}{n} = \frac{18}{100}$$
$$4300 = 18n$$
$$238.\overline{8} = n$$

23. The interest I is $I = 8000(0.06)(0.75)$ and the account balance after 9 months is the original $8000 plus the
$$I = \$360$$
interest of $360 or $8360.

25. To compute 40% of $7000 mentally you could first consider a "simpler" question, "what is 10% of $7000?" 10% is found by a decimal shift one unit to the left, $700, so 40% is four times this amount or $2800.

27. a) The percent increase is found by asking 30 is what percent of 80? $30 = \dfrac{n}{100} \times 80$ so the dress increased
$$3000 = 80n$$
$$37.5 = n$$

   in price by 37%.

   b) $$\frac{30}{80} = \frac{n}{100}$$
$$\frac{3000}{80} = n$$
$$37.5 = n$$

29. The sales tax is found by asking $0.57 is what percent of $9.49?

$$\frac{0.57}{9.49} = \frac{n}{100}$$

$$\frac{57}{9.49} = n$$

$$6\% \approx n$$

31. a) $0.14379\ldots$ is irrational

b) $\sqrt{7}$ is irrational

c) $0.\overline{23}$ is rational

d) 200% is rational

33. $\sqrt{17}$ is (d) an irrational number and (e) a real number.

35. The set of real numbers is the union of the set of rational numbers and the set of irrational numbers. The set of rational numbers contains all terminating and repeating decimals, while the set of irrational numbers contains all non – repeating decimals.

37. A counter-example that real number division is not associative could be $5 \div (10 \div 2) \neq (5 \div 10) \div 2$

$$5 \div 5 \neq \frac{1}{2} \div 2$$

$$1 \neq \frac{1}{4}$$

Chapter Eight

Section 8.1

LE 1   Opener

a) Three things in your classroom may include a pencil as a line segment, a sheet of paper as a plane, and a

clock as a circle

b) The chalkboard surface suggests a plane, a corner (the intersection of two walls and the floor) suggest a

point, and the intersection of two walls suggest a line segment

c) Compare your thoughts with your peers

LE 2   Concept

A geometric line never ends whereas a drawn line does

LE 3   Reasoning

If A and B are points in a plane, then $\overleftrightarrow{AB}$ lies entirely in the plane is a true statement

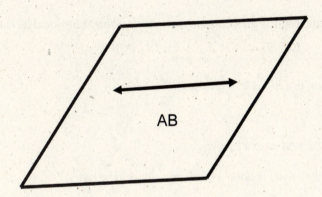

LE 4   Concept

a) A very thin string has dimension one

b) A gorilla has dimension three

c) The surface of a table has dimension two

LE 5   Reasoning

A line segment is a subset of a line bounded by two endpoints. A ray is a subset of a line that is bounded at

one point and extends infinitely in the other direction

LE 6   Reasoning

An angle is the union of two rays sharing a common vertex

LE 7   Reasoning

a) $\overline{ME} \cup \overline{ET}$ is $\overline{MT}$

b) $\overline{NU} \cap \overline{CE}$ is $\overline{CN}$

c) $\overline{ME} \cap \overline{CN}$ is ϕ

LE 8   Skill

a) The measure of $\angle SUN$ is approximately 70^0

b) With a protractor the measure of $\angle SUN$ is 70^0

LE 9   Reasoning

The fourth grader needs to recognize if she is measuring an acute angle (angle measure less than ninety

degrees) or an obtuse angle (angle measure greater than ninety degrees). If the angle is less than ninety

degrees than 60 would be correct and if the angle is greater than ninety degrees than 120 would be correct.

LE 10   Connection

The most common angles are right angles, that is angle of measure 90^0

LE 11   Connection

a) An object that illustrates this property are doors of a cabinet

b) The property enables them to fit together and the bottom of the two doors to be level

LE 12   Connection

An object in the classroom that forms an acute angle could be a partially opened door

LE 13   Reasoning

a) $\angle ACE$ and $\angle ACT$ are supplementary angles

b) Using a definition to answer part (a) involves deductive reasoning

LE 14   Concept

a) All acute angles are congruent is a false statement

b) 34^0 and 20^0 are both acute angles but not congruent

LE 15   Reasoning

a) Two intersecting lines have exactly one point in common

b) Two lines are perpendicular if and only if they intersect at right angles

c) Two parallel lines lie in the same plane and do not intersect

LE 16   Reasoning

a) In a plane, two lines that are both parallel to a third line must be parallel to each other is a true statement

b) An example might be the chalkboard. In my classroom the chalkboard has a frame with a middle

separator, thus the two end vertical frame pieces are parallel to the vertical center piece.

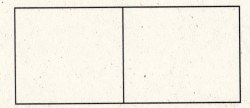

LE 17   Summary

Lines continue indefinitely in a plane. Angles can be acute in measure, obtuse in measure, or right angles.

Two angles whose measures sum to ninety degrees are complementary angles. Two angles who measures

sum to 180^0 are supplementary angles.

Section 8.1 Homework Exercises

1.   The root geo means earth and metry means measure

3.   a) Three objects at home that suggest common space figures could be a television (rectangular prism), a

ball (sphere), and drinking class (cylindrical)

b) Three objects at home that suggest common plane figures could be a shelf, the floor, or the ceiling

5. a) This is a photograph of the Pentagon and suggest a pentagon

    b) A special piece of glass that disperses light into its color components suggests a prism

    c) The edge of a box suggest a line segment

7. Euclid lived about (d) 2000 years ago

9. A geometric point is different from a dot in that the point has no length or width

11. a) If a line $\overleftrightarrow{AB}$ intersects a point in plane c, then $\overleftrightarrow{AB}$ lies in plane c is a false statement.

    b)

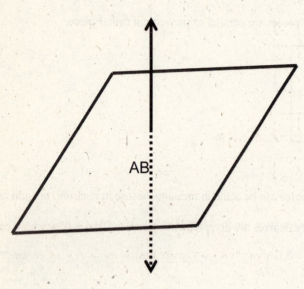

13. a) A pillow is three dimensional

    b) A very thin piece of wire no matter what way we see it is three dimensional, because that is the human

    eye experience. Perhaps a thin piece of wire having no thickness and/or placed on a two dimensional plane

    would be viewed as one dimensional

    c) A sheet of paper is two dimensional

15. a) A line segment has two endpoints; a line has no endpoints

    b) A line segment consists of two points on a line and all the points between, when we draw a line segment

    we don't sketch all the points, we use a pencil to connect the two endpoints.

17. a) The vertex of $\angle PAT$ is A

    b) $\overrightarrow{AP}$ is a ray with endpoint A going in the direction of point P, whereas $\overrightarrow{PA}$ is a ray with endpoint P going

    in the direction of point A

    c) Rays $\overrightarrow{AP}$ and $\overrightarrow{AT}$ are the sides of $\angle PAT$

19.   a) $\overrightarrow{AB} \cup \overrightarrow{AC}$ is $\angle ABC$

b) $\overrightarrow{AB} \cup \overrightarrow{AD}$ is $\overline{BD}$

c) $\overrightarrow{BA} \cap \overrightarrow{AC}$ is A

21.   a) This angle is approximately 120 degrees, when measured it is 121 degrees

b) The scissors measure is approximately 30 degrees, when measured it is 32 degrees

23.

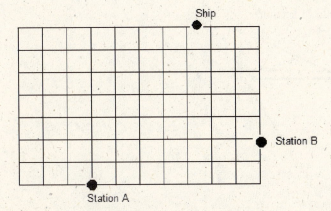

25.   a) The child might think that angle A is larger since the rays representing the angle are longer.

b) You could place angle B on top of angle A (using a transparency reproduction) to show that the opening

is wider.

27.   Two capital letters that contain two or more acute angles and no obtuse or right angles could be two of the

following W, M, Z, N.

29.   a) Two supplementary angles are $\angle PIN$ and $\angle PIG$

b) Two complementary angles are $\angle PIR$ and $\angle RIG$

c) Two more supplementary angles are $\angle RIN$ and $\angle RIG$

31.   Two complementary angles are congruent is a false statement

33. $\overline{FG}$ is congruent to $\overline{AB}$

35.   a) True statement

b) Two parallel lines on the floor of a room that are perpendicular to a third line

c) A carpenter could tell that two lines are parallel if they both form a right angle with a straightedge tool.

37.   This is correct

39.   a) Four lines that separate the plane into 10 sections.

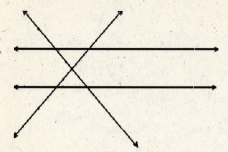

b) Four lines that separate the plane into 11 sections.

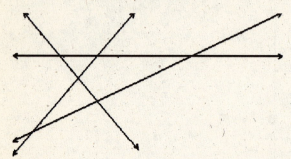

41.   a)

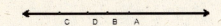

b)

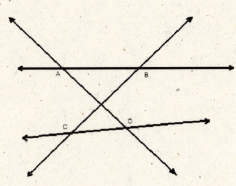

c)

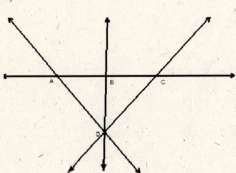

43.   a) From point A building 1 can be seen.

b) From point B building 1 can be seen.

c) From point C buildings 1 and 3 can be seen.

d)

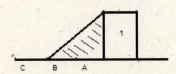

45.   a) 5 a.m. and 7 a.m.

b) 2:30 a.m. and 9:30 a.m.

47.

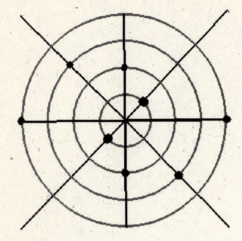

49.   For student to read.

Section 8.2

LE 1   Opener

A polygon is a simple closed plane figure bounded by line segments.

LE 2   Reasoning

A polygon is a simple closed plane figure formed by three or more line segments.

LE 3   Concept

A polygon must be closed, have the same starting and stopping point, and be made up of line segments.

LE 4   Concept

a) This is a pentagon.

b) This is a quadrilateral.

LE 5   Reasoning

a)

Polygons				
Number of Vertices	3	4	5	6
Number of diagonals from each vertex	0	1	2	3
Total number of diagonals	0	2	5	9

b) In a heptagon there should be 14 diagonals.

c)

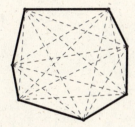

LE 6   Reasoning

a) An N gon has N vertices.

b) $N-3$ diagonals can be drawn from each vertex.

c) The total number of diagonals is the product of the number of diagonals from one vertex and the number of vertices divided by 2.

d) The polygon has $\dfrac{N(N-3)}{2}$

LE 7   Reasoning

a) Let $N = 10$,
$$\frac{N(N-3)}{2} = \frac{10(10-3)}{2}$$
$$= \frac{10(7)}{2}$$
$$= 35$$

b) The process of assuming that the formula is true and applying it in part (a) involves deductive reasoning.

LE 8   Summary

A polygon must be closed, have the same starting and stopping point, and be made up of line segments.

Section 8.2 Homework Exercises

1.   a) This is not a polygon as it is not a plane figure.

b) This is a polygon.

c) This is not a polygon, it is not simple.

d) This is not a polygon as the sides are not segments.

3.   a) The stop sign suggests an octagon.

b) A trapezoid is suggested in this picture.

5.

7.   a)

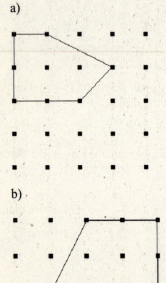

b)

9.   Choices (b) and (c) are convex polygons.

11.   a)

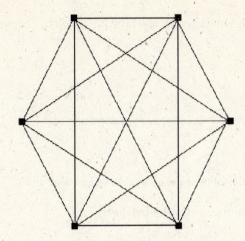

       b)   There are 9 diagonals.

13.   a) The maximum number of intersections points for a triangle and a square is 6.

       b) The maximum number of intersections points for a triangle and a convex pentagon is 6.

       c) The maximum number of intersections points for a square and a convex pentagon is 8.

       d) The maximum number of intersections points for a regular polygon with n sides and a regular polygon

       that has m sides is $2n$.

15. $$\frac{15(15-3)}{2}+15=90+15$$
       $$=105$$

       This is the sum of the number of diagonals and the number of "sides" if you were to construct a polygon.

17.   For student

Section 8.3

LE 1   Connection

       Triangles are more stable because the three vertices of a triangle can always be contained in one plane.

LE 2   Opener

       a) Triangles could be grouped by the number of congruent sides or angles.

       b) If triangles are grouped by congruent sides there could be: equilateral, scalene, isosceles.

       If triangles are grouped by angles there could be: right, acute, obtuse.

LE 3   Concept

    a) This is an obtuse scalene triangle.

    b) This is an acute equilateral triangle.

LE 4   Concept

Parallelograms, rectangles, rhombi, and squares have both pairs of opposite sides congruent.

LE 5   Reasoning

    a) All parallelograms and rectangles share the properties of opposite congruent and parallel sides.

    b) All rectangles have interior right angles, but this is not necessary for parallelograms.

LE 6   Reasoning

    a) All sides are congruent for squares and also for rhombi.

    b) All squares have interior right angles, this is not necessary for rhombi.

LE 7   Concept

Trapezoid – is a quadrilateral with exactly one pair of parallel sides.

Parallelogram – is a quadrilateral in which opposite sides are parallel and congruent.

Rectangle – is a parallelogram with at least one right angle.

Rhombus – is a parallelogram with four congruent sides.

Square – is a rhombus with four right angles.

LE 8   Reasoning

The minimum number of right angles that makes a parallelogram a rectangle is one.

LE 9   Reasoning

    a) The second child is correct.

    b) State the definition to the child and ask him/her to check if a square has all the requirements.

LE 10   Reasoning

    a) No, every trapezoid is not a parallelogram. A parallelogram is a quadrilateral in which both pairs of

    opposite sides are parallel, whereas a trapezoid has exactly one pair of parallel sides

    b) Yes, every square is also a rhombus. A square is a parallelogram that has four congruent sides and four

    right angles, this satisfies the rhombus condition of four congruent sides.

LE 11   Reasoning

   a) $Rh \bigcap Re = S$

   b) $T \bigcap P = \phi$

LE 12   Reasoning

   a)

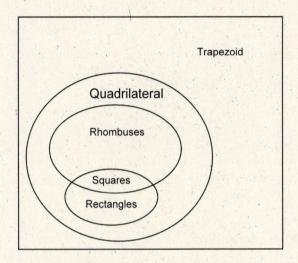

   b) Choice (ii)

LE 13   Concept

a)

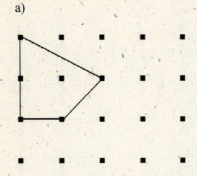

b) Impossible

c)

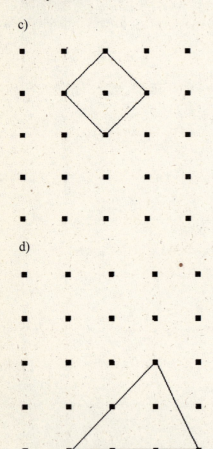

d)

LE 14   Concept

A chord is a line segment with endpoints on the circle.

LE 15   Concept

a) A diameter is a chord that passes through the center of a circle.

b) You could describe a circular object by its area, circumference, diameter, or radius.

LE 16   Reasoning

Each minute mark on a clock is $\frac{1}{60}$ of 360^0 or 6 degrees. 12.5 minutes past the hour hand would constitute

75^0, so if the hour hand were on 3, then the minute hand at $27\frac{1}{2}$ past 3 o'clock would create an example

of 75^0.

LE 17   Reasoning

Opposite angles of a parallelogram are congruent.

LE 18   Summary

Triangles are classified according to their angle measures and sides. Triangles studied include scalene,

equilateral, and isosceles. Quadrilaterals are classified according to whether opposite sides are parallel or

not and also whether they have right angles or not. Quadrilaterals studied include the trapezoid, rectangle,

parallelogram and square.

Homework Exercises 8.3

1.   a) Isosceles, acute

b) Isosceles, acute

c) Scalene, obtuse

3. a)

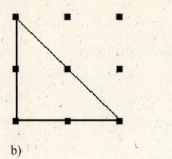

b)

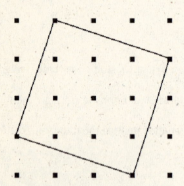

5.

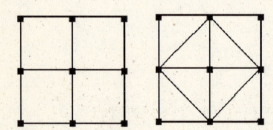

7. Check the angles to verify they are right angles.

9.

11. a) For a parallelogram you could use 4 from group 1 or 2 from group 1 and 2 from group 2.

b) For a trapezoid you could use 1 from group 3 with 1 or 2 from group and 2 or 1 from group 2.

c) A rhombus you could use 4 from group 2.

13. These are all true.

15.   a)

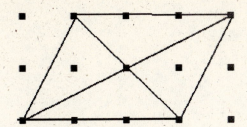 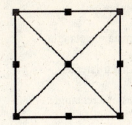

    b) The diagonals bisect each other.

    c) Diagonals of a rhombus are perpendicular.

17.   a) Rhombus can have at least one right angle.

    b) Square must have at least one right angle.

    c) Trapezoid can have at least one right angle.

    d) Rectangle must have at least one right angle.

    e) Parallelogram can have at least one right angle.

19.   They are all quadrilaterals with opposite sides parallel and congruent.

21.   Rectangle, parallelogram, pentagon, hexagon, triangle, kite, heptagon are possible shapes.

23.   a) Correct

    b) Correct

    c) Incorrect

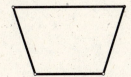

25.   Three is the minimum number of right angles that makes a quadrilateral a rectangle.

27. A square is a special case of a rectangle

29. a) All squares are rhombuses

   b) Some parallelograms are rectangles

   c) All trapezoids are quadrilaterals

31. It has the all properties of a rectangle.

33. A square is a rectangle.

35. Set b)

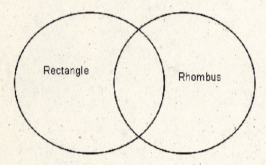

37. a) $P \bigcap S = S$

   b) Yes

   c) $P \bigcup Q = Q$

39. The order could be b), a), d), then c)

41. Circle N is the set of points in a plane that are 8 inches from N.

43. a) $m\angle DCF = 90^0$

   b) This is an isosceles triangle.

   c) $\overline{BG}$

45.

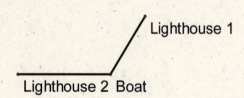

47.   a) – c)

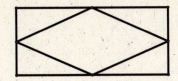

      d) A parallelogram

      g) Induction

49.   a) is the correct Venn diagram.

51.   a)

      b)

      c)

      d)

53.   a) – c)

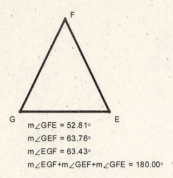

m∠GFE = 52.81°
m∠GEF = 63.76°
m∠EGF = 63.43°
m∠EGF+m∠GEF+m∠GFE = 180.00°

e) The sum of the angle measures of a triangle is 180^0.

55.   For the student

57.   For the student

Section 8.4

LE 1   Opener

a) Sum of the angle measures of a triangle is 180°.

b) Sum of the angle measures of a quadrilateral is 360°.

LE 2   Reasoning

The sum of the 3 angles in a triangle forms a straight angle which is equal to 180°.

LE 3   Reasoning

$180 \div 3 = 60^o$ for each angle measure

LE 4   Reasoning

The sum of the measures would exceed 180°.

LE 5   Reasoning

a) 180^0

b) 360^0

LE 6   Reasoning

a) Together the angles of the four triangles do not form the four angles of the quadrilateral.

b) $(4 \bullet 180^o) - 360^o = 360^o$

LE 7   Reasoning

3 triangles are formed, $3 \bullet 180^o = 540^o$, the sum of the interior angle measures of a pentagon.

LE 8   Reasoning

a)

Polygons (convex)						
Number of sides	3	4	5	6	7	N
Number of Triangles formed	1	2	3	4	5	$N-2$
Sum of interior angle measures	180	2(180)	3(180)	4(180)	5(180)	$(N-2)180$

b) The sum of the interior angle measures of any polygon is given by $(N-2)180^0$ where N represents the number of sides of the polygon.

LE 9   Skill

$(8-2) \bullet 180^o = 1080^o$

LE 10   Skill

$(8-2) \bullet 180^o = 1080^o$ is the sum of the interior angles of the regular octagon, stop sign. Each angle has

measure $\dfrac{1080^o}{8} = 135^o$

LE 11   Skill

In Figure 8-39 the angle measures around the marked point add up to 360 degrees.

LE 12   Reasoning

a) Equilateral triangles tessellate.

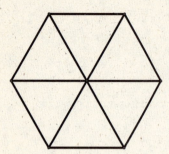

b) The measure of each angle of an equilateral triangle is 60^0. 60 is a factor of 360, the sum of the angles

about a point.

LE 13   Reasoning

a) A regular pentagon does not tessellate the plane.

b) Each angle in a regular pentagon has measure $(5-2) \bullet 180^o = 540^o$, $\dfrac{540^o}{5} = 108^o$. 108 is not a factor of

360, so a regular pentagon does not tessellate.

LE 14   Reasoning

a)

Regular Polygons	Total Number of Degrees	Measure of Each Interior Angle
Triangle	180	60
Square	$(4-2) \bullet 180^o = 360^o$	90
Pentagon	$(5-2) \bullet 180^o = 540^o$	$\frac{540^o}{5} = 108^o$
Hexagon	$(6-2) \bullet 180^o = 720^o$	$\frac{720^o}{6} = 120^o$
Heptagon	$(7-2) \bullet 180^o = 900^o$	$\frac{900^o}{7} \approx 128.57^o$
Octagon	$(8-2) \bullet 180^o = 1080^o$	$\frac{1080^o}{8} = 135^o$
Decagon	$(10-2) \bullet 180^o = 1440^o$	$\frac{1440^o}{10} = 144^o$
Dodecagon	$(12-2) \bullet 180^o = 1800^o$	$\frac{1800^o}{12} = 150^o$

b) Triangles, quadrilaterals, and hexagons all tessellate.

LE 15   Summary

The angle measures of a polygon with n sides is $(n-2) \bullet 180^o$. The measures of the interior angles of a regular polygon are congruent. Polygons whose interior angle measure will divide 360 tessellate the plane.

1.   a) – b)

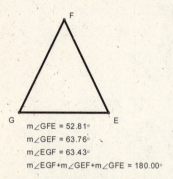

m∠GFE = 52.81°
m∠GEF = 63.76°
m∠EGF = 63.43°
m∠EGF+m∠GEF+m∠GFE = 180.00°

c) The sum of the angles should be close to 180 degrees, it may not be exact due to drawing and/or measuring error.

3.   A triangle could not have two right angles as their sum would be 180^0 and the sum of the 3 angle measures should equal 180^0.

5.   Since triangle KAR is equilateral each interior angle is 60^0. The measure of angle AKP is $108 - 60 = 48^0$. The measure of angle KAP is $180 - (110 + 48) = 22^0$.

7.   The measures of the other two angles is 180^0.

9.

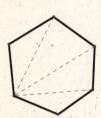

The sum of the angle measures of the 4 triangles is $4 \bullet 180 = 720^0$. Therefore the sum of the angle measures of the hexagon is also 720^0.

11.   The angles of the five triangles do not form the angles of the pentagon.

13. $(40 - 2) \bullet 180^0 = 6840^0$

15.   a) $\frac{(6-2) \bullet 180^0}{6} = 120^0$

b)

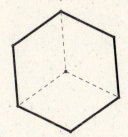

17.   a)

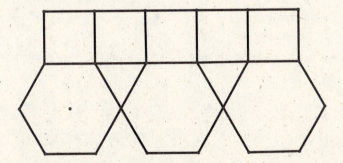

b) The other six are: 2 octagons and 1 square; 1 square, 1 hexagon, and 1 dodecagon; 2 squares, 3 triangles;

2 squares, 3 triangles (a different way); 1 hexagon and 4 triangles; 2 dodecagons and 1 triangle.

19.   a) A parallelogram tessellates the plane.

b) A trapezoid tessellates the plane.

21.   a) There are 6 rectangles.

b) There are 10 rectangles.

c) There are 15 rectangles.

d) $\frac{n(n+1)}{2}$ for n rectangles

e) Induction

f) 36

23.   a) The measure of each interior angle is $\frac{(n-2)180^0}{n}$

b) Because $\dfrac{(n-2)}{n}$ is less than 1.

25.

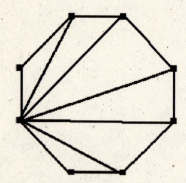

$\dfrac{n(n-3)}{2} = \dfrac{8 \bullet 5}{2}$
$\qquad\qquad = 20$

$20 + 8 = 28$ matches

27.　　For the student

29.	a)

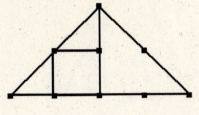

b)

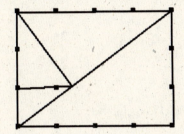

c)

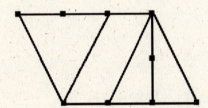

d)

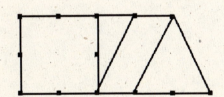

Lesson Exercises 8.5

LE1	Opener

They can be skew, they never intersect and are not in the same plane.

LE2	Reasoning

a) False

b) An example could be the line that forms the intersection of the front wall and ceiling is skew to the line that forms the intersection of a side wall and the floor, but parallel to the line that forms the intersection of the back wall and the ceiling.

LE3   Connection

A chandelier hanging from the ceiling.

LE4   Concept

a) True

b) The legs of a table are perpendicular to the floor and parallel to each other.

LE5   Opener

The two planes can either intersect or be parallel.

LE6   Reasoning

a) True

b) Room dividers that are each parallel to the wall are parallel to each other.

LE7   Concept

a) Triangular pyramid.

b) Rectangular prism.

LE8   Concept

A polyhedron is a simple closed space figure bounded by polygonal regions.

LE9   Concept

a) A cube has 6 faces.

b) A cube has 8 vertices

c) A cube has 12 edges.

LE10   Communication

The method the student has used will count the same edge twice, the total would be $24 \div 2 = 12$.

LE11   Reasoning

A prism has 2 opposite faces that are parallel, pyramid, spheres, and cones does not.

LE12   Concept

a) A triangular prism has 5 faces.

b) A triangular prism has 6 vertices.

c) A triangular prism has 9 edges.

LE13   Reasoning

a)

Figure	Faces F	Vertices V	Edges E
Cube	6	8	12
Triangular Prism	5	6	9
Rectangular Prism	6	8	12
Pentagonal Prism	7	10	15

b) $F + V - E = 2$

LE14   Reasoning

a)

Figure	Faces F	Vertices V	Edges E
Cube	6	8	12
Triangular Prism	5	6	9
Rectangular Prism	6	8	12
Pentagonal Prism	7	10	15
Triangular Pyramid	4	4	6
Square Pyramid	5	5	8

b) $E = F + V - 2$

LE15   Concept

a)

Regular Polyhedron	Faces F	Vertices V	Edges E
Cube	6	8	12
Tetrahedron	4	4	6
Octahedron	8	6	12
Icosahedron	20	12	30
Dodecahedron	12	20	30

b) Yes, Euler's formula works.

LE16   Concept

a) Three

b) The angle measures would total 360^0.

c) 3; 4; 5

d) The angle measures would total 360^0.

e) Three

LE17   Concept

Cylinders, cones, and spheres are not polyhedra as their faces are not polygonal regions.

LE18   Concept

You can construct the lateral surface of a right circular cylinder from a rectangle.

LE19   Concept

a) All cones and pyramids have only one base.

b) Analysis

LE20   Connection

a) A square prism can be made with this net.

b) A pentagonal pyramid can be made with this net.

LE21   Concept

a) Hexagonal Prism

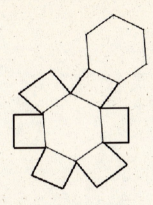

b) Square Pyramid

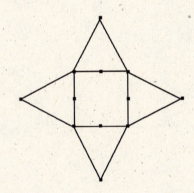

LE22   Reasoning

Figures a, c, e, l, and p are nets for cubes.

LE23   Summary

Polyhedra are simple closed space figures whose faces are polygons. They adhere to Euler's formula

$$F + V - E = 2.$$

Homework Exercises 8.5

1.   a) False

b) The three lines segments that meet at a corner of a room.

3.   a) Infinite

b) 1

c) 1

d) An infinite number.

5.	False

7.	True

9.	a) An infinite number of planes pass through *A*.

	b) An infinite number.

	c) 1 or infinite number

	d) For student

11.	a) Two skew lines are AE and FG.

	b) The plane that contains ABEF is parallel to the plane that contains DCGH.

	c) The plane that contains ABCD is perpendicular to the plane that contains BCGF.

	d) The intersection is line FG.

13.	True

15.	a) A picture frame suggests polygonal region.

	b) A page in this book suggests a polygon.

	c) A face of a triangular prism suggests a polygon.

17.	a and d are polyhedra.

19.	a) A right circular cylinder.

	b) A rectangular prism.

21.	a) EGCA is a rectangle.

	b) Triangle EGH is isosceles.

23.	This solid figure would have 13 vertices

25.	Prisms have parallel bases, pyramids do not.

27.	a) A hexagonal prism has 8 faces.

	b) A hexagonal prism has 18 edges.

	c) A hexagonal prism has 12 vertices.

29.	A pentagonal pyramid has 6 vertices and 6 faces.

31.   a) $F + V - E = 2$ True
      $9 + 9 - 16 = 2$

      b) $F + V - E = 2$ True
      $7 + 10 - 15 = 2$

      c) $F + V - E = 2$ True
      $18 + 14 - 30 = 2$

33.   a) A cylinder and a cone.

      b) 2 cylinders.

35.   They might respond that they "see" a drum.

37.   A pentagonal prism.

39.   A sphere has no thickness.

41.   a) The net could make a pentagonal prism.

      b) The net could make hexagonal pyramid.

      c) The net could make a hexagonal prism.

43.   a)

      b)

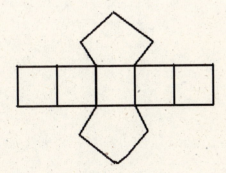

45.   a, c, and d fold into cubes.

47.   a) Parallelogram.

      b) A cylinder holds its shape more firmly.

49.   a) $n+2$

   b) $2n$

   c) $3n$

   d) Yes, Euler's formula works for prisms.

51.   a) Octahedron.

   b) Tetrahedron.

   c) Cube.

   d) y vertices, x faces, and z edges.

53.   c) The line on the regular loop is on one side. The line on the Mobius strip goes all around it on both

   "sides." It has only one side.

Lesson Exercises 8.6

LE1   Concept

   Part (a) is true.

LE2   Concept

   No.

LE3   Opener

   a)

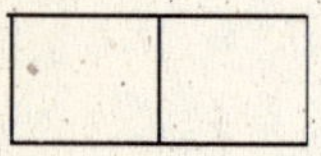

   b)

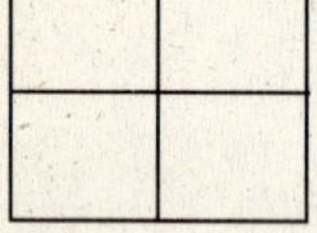

   c)

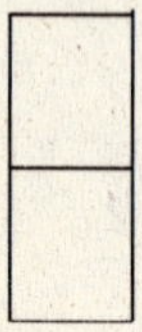

LE4   Concept

a)

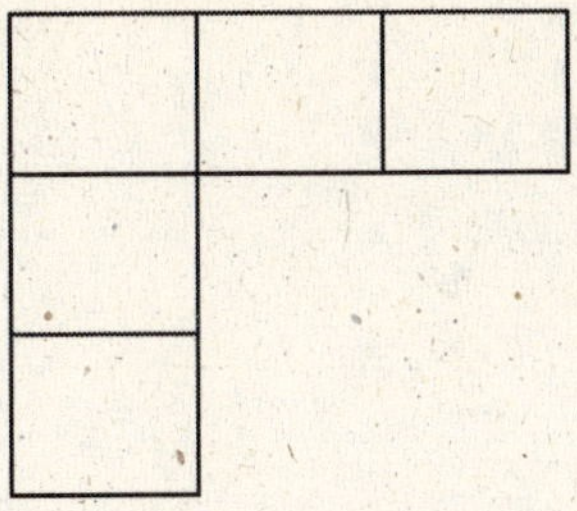

Base Outline

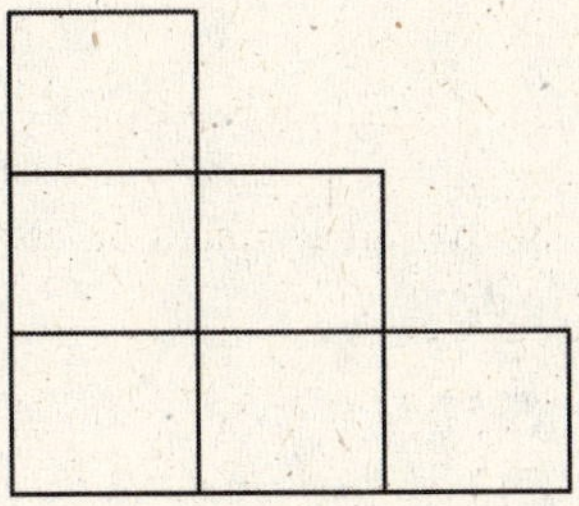

Front

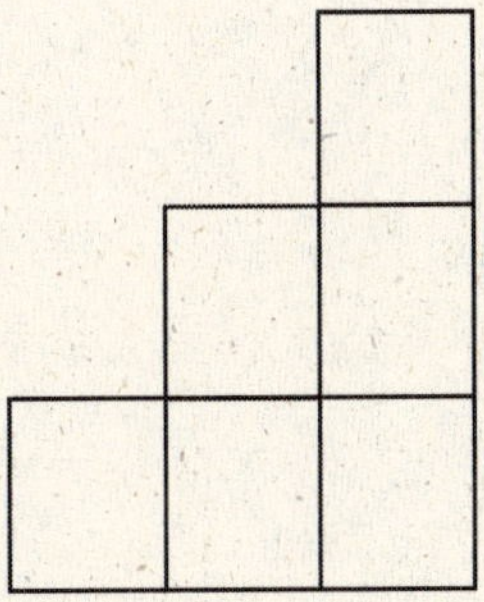

Right Side

b)

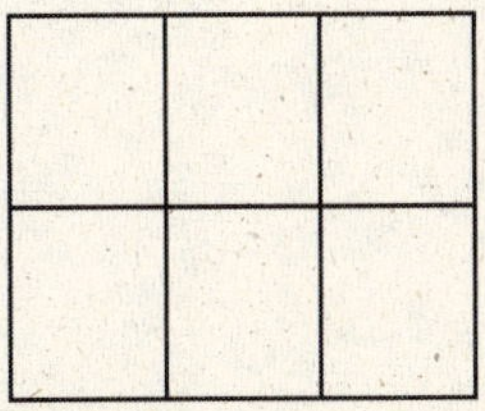

Base Outline

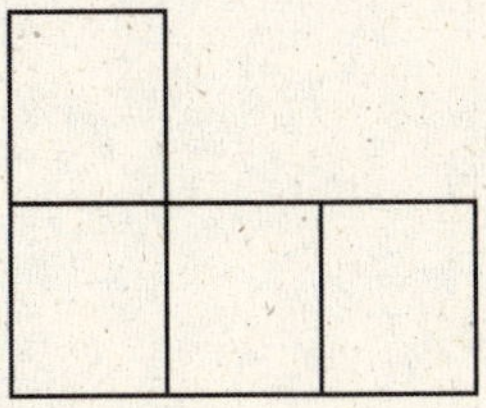

Front

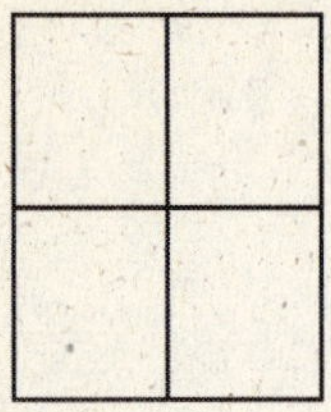

Right Side

LE5   Concept

2	1	1
2	1	1

LE6   Skill

a)

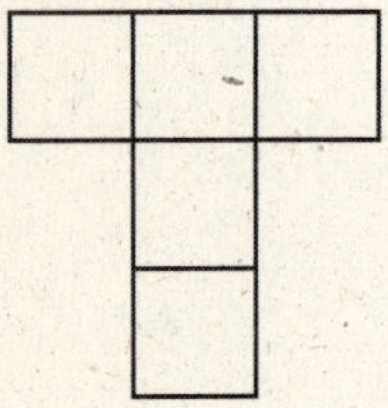

Top

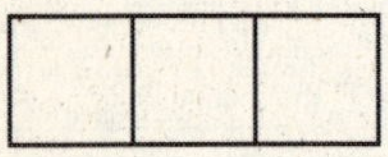

Front

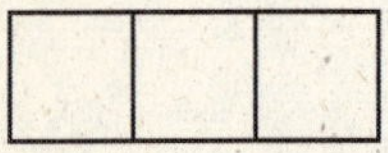

Right

b)

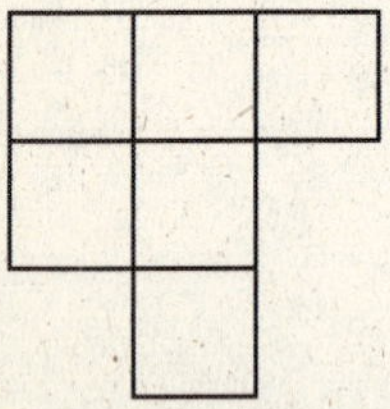

Top

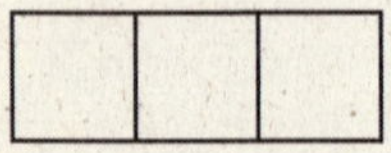

Front

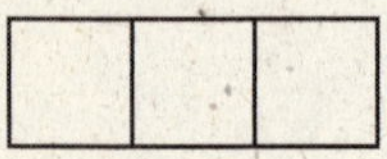

Right

LE7   Skill

LE8   Reasoning

Triangular prism.

LE9   Skill

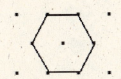

LE10   Opener

a)

b) 120^0

LE11   Skill

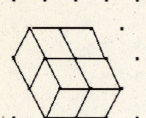

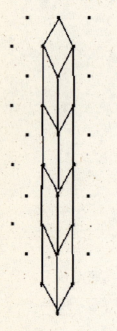

LE12   Concept

a)

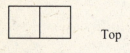   Top

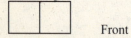   Front

   Right Side

LE13   Concept

All the furniture looks crooked.

LE14   Concept

Lines going from left to right across the floor or ceiling are drawn parallel and do not pass through the vanishing point. Only parallel lines from the front to back are drawn so that they meet at the vanishing point.

LE15   Skill

Follow steps in textbook.

LE16   Skill

Vanishing Point

LE17   Summary

Perspective drawings are informative for understanding views of structures. Vanishing points are very helpful in drawing perspective views.

Exercises 8.6

1.   a) AB appears to be longer than CD.

        b) CD is longer than AB.

        c) We tend to look for both depth, a third dimension.

3.   The room is not rectangular. The back of the room on the right side is closer to the camera.

5.   a)

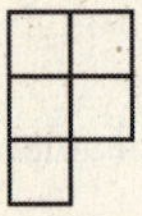

Base Outline

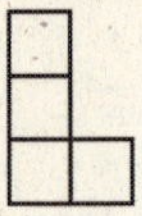

Front

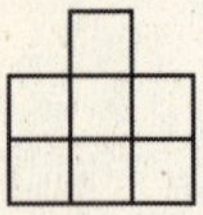

Right Side

b)

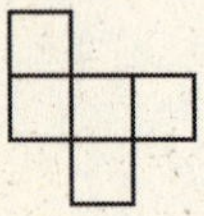

Base Outline

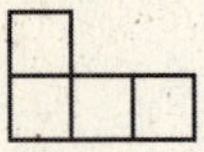

Front

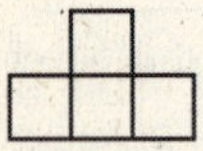

Right

7.   a)

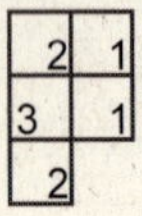

b)

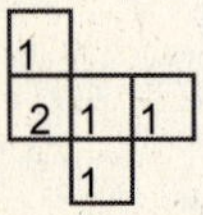

9.

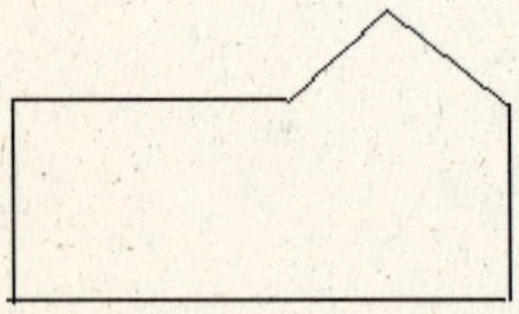

Front

Top

11. a)

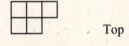 Top

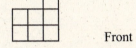 Front

 Right

b)

Top

Front

Right

13.

	1	3
2	2	
1	2	1

15.   The highest stack on the front view is 3, highest on the right side is 4.

17.   Hexagonal prism.

19.   a) The resulting cross section is a triangle.

b) The resulting cross section is a rectangle.

21.   Square pyramid and square prism.

23.

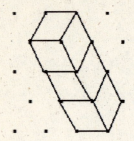

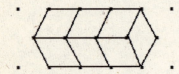

25.   a)

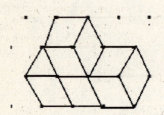

b)

   Base Outline

Front

Right

27.

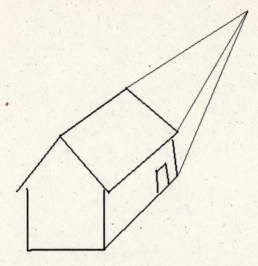

29.   Draw your last name using the steps for block lettering on page 459.

31.   The fisherman in the bottom right reaches the water with his pole. The man in the upper right on the hill

lights the women's lantern. The man on one side of the bridge shoots his rifle on the other side. The banner

from the building hangs out among the trees.

33.

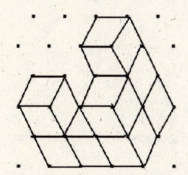

35.

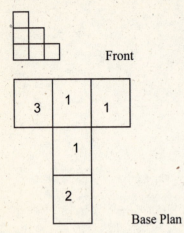

Front

Base Plan

37.   a) In this order, front, top and side.

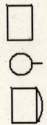

b) In this order, front, top and side.

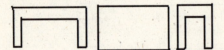

39.   a) 64

b) The 4 center faces of the 4 center cubes on each of the 6 larger cube faces, so a total of 24 cubes would

have exactly one face painted.

c) The 2 center cubes on each edge have exactly 2 faces painted and there is a total of 16 of these.

d) Each corner cube would have 3 faces painted, a total of 8.

e) There are no cubes with 4 faces painted.

f) There are a total of 64 smaller cubes 48 of which have been painted, $64 - 48 = 16$ cubes without paint.

41.   a) 1 face painted: $16 \times 6 = 96$

2 faces painted: $4 \times 8 = 32$

3 faces painted: 8

- 4 faces painted: None

0 faces painted: $216 - (96 + 32 + 8) = 80$

b) 1 face painted: $(n-2)^2 \times 6 = 6n^2 - 24n + 24$

2 faces painted: $(n-2) \times 8 = 8n - 16$

3 faces painted: 8

4 faces painted: None

0 faces painted: $n^3 - (6n^2 - 24n + 24) - (8n - 16) - 8 = n^3 - 6n^2 + 16n - 16$

Chapter 8 Review Exercises

1.   To define space figures properly, one must first study the components of these figures.

3.   a) $\angle ABC$

b) {}

5.   a) $\angle ABC = 26^0$

b) Acute

c) The complement is $90^0 - 26^0 = 64^0$

7.

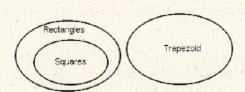

9.   a) Some parallelograms are rhombuses.

b) All squares are rectangles.

11.

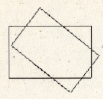

13.   There are 3 triangles inside the pentagon. Sum of the interior angles in each triangle is 180^0.

$$3 \times 180^0 = 540^0$$

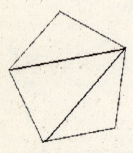

15.   All of the interior angles of the pentagon are 108^0. $\angle FBC = \angle FCB = 72^0$, $\angle BFC = 36^0$

17.   Yes, each angle is 120^0, $3 \times 120^0 = 360^0$

19.   Infinite number of planes.

21.   False.

23. $F + V - E = 2$
       $F + 14 - 21 = 2$
       $\quad\quad F = 9$

25.

      Front

      Right

Chapter Nine

Lesson Exercises 9.1

LE1   Opener

a) Translation (slide) down 2

b) Reflection

c) Rotation clockwise 90^0

d) Glide (reflection and translation)

LE2   Skill

a)

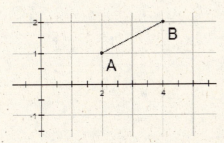

b)

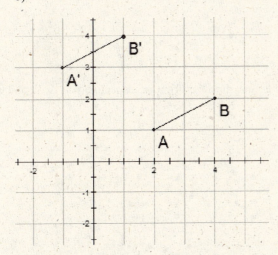

c) $A(2,1) \rightarrow A'(-1,3) \ B(4,2) \rightarrow B'(1,4)$

d) $(a,b) \rightarrow A'(a-3,b+2)$

LE3   Concept

a) One triangle looks like a mirror reflection of the other when the line is used as a mirror.

b) Use a mirror to verify reflection.

LE4   Reasoning

a) - d)

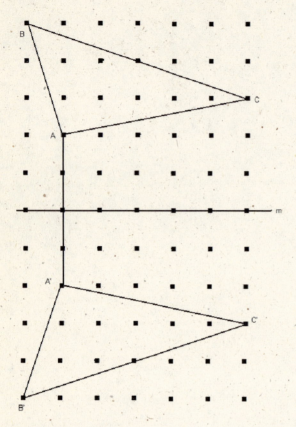

e) $\overline{AA'}$ is perpendicular to line *m*.

f) A', B', and C' are on the opposite side of *m*, the same distances from *m*, respectively as A, B, and C.

LE5   Skill

a), b)

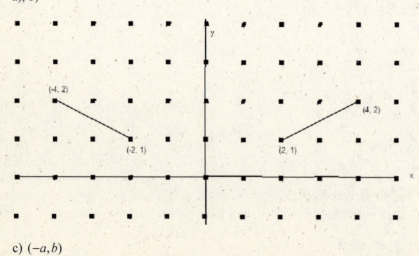

c) $(-a, b)$

LE6   Skill

a), b)

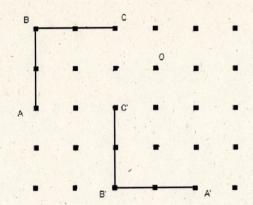

LE7   Concept

The shape was rotated 360^0.

LE8   Concept

The shape was rotated 180^0.

LE9   Skill

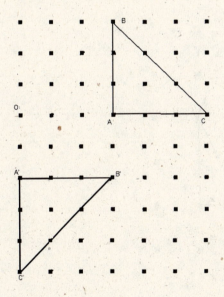

LE10   Concept

a) Yes.

b) A rotation of 360^0 has the same property.

LE11 Skill

a), b)

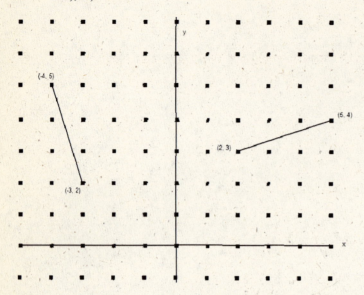

c) $E(2,3) \rightarrow E'(-3,2)$ $F(5,4) \rightarrow F'(-4,5)$

d) $(-b,a)$

LE12 Reasoning

a), b)

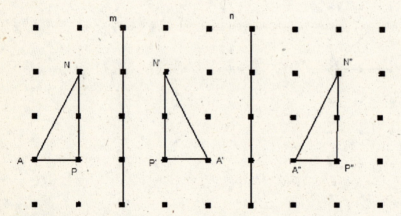

c) Translation

d) It is twice as much.

LE13 Reasoning

a) They have the same size and shape.

b) Its position is different.

LE14   Concept

Yes

LE15   Concept

Parts (a) and (c) contain congruent figures by a translation and a reflection respectively.

LE16   Opener

Part (d) needs a reflection and a translation.

LE17   Skill

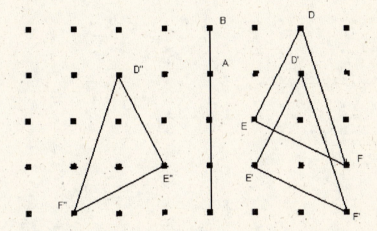

LE18   Summary

The rigid motions of translation (slide), rotation, and reflection can be used to show congruence between

shapes.

Homework Exercise 9.1

1.

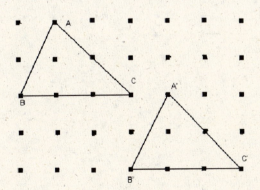

3.　　a), b)

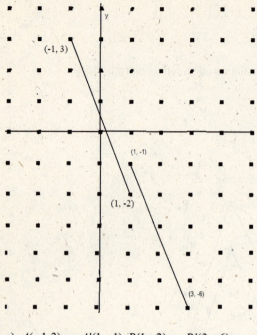

(-1, 3)

(1, -1)

(1, -2)

(3, -6)

c) $A(-1,3) \rightarrow A'(1,-1) \ B(1,-2) \rightarrow B'(3,-6)$

d) $(a+2, b-4)$

5.   a), c)

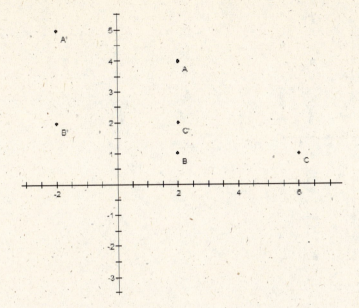

b) $A'(-2,5), B'(-2,2), C'(2,2)$

d) Translation

7.

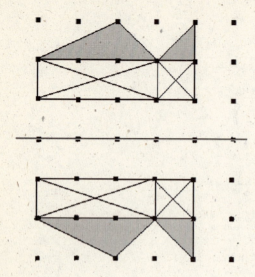

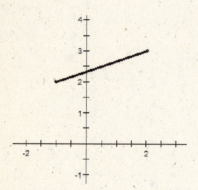

9.   a)

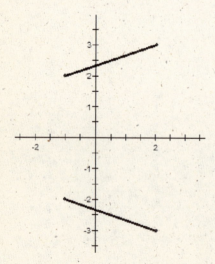

b)

c) $A(-1, 2) \rightarrow A'(-1, -2)$ $B(2, 3) \rightarrow B'(2, -3)$

d) $(a,-b)$

11.

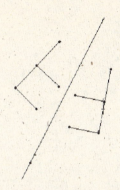

13.   Fold the paper so that the shape and its image coincide. The folding line will be the line of reflection.

15.   When this is reflected in a vehicles rear view mirror they can read, AMBULANCE.

17.   a) – d)

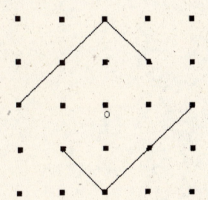

19.   a), b)

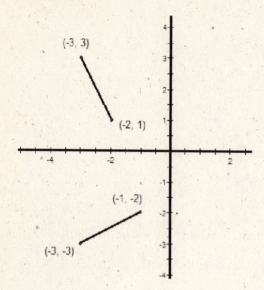

(-3, 3)

(-2, 1)

(-1, -2)

(-3, -3)

c) $A(-2,1) \rightarrow A'(-1,-2)$ $B(-3,3) \rightarrow B'(-3,-3)$

d) $(-b, a)$

21.   a), b), c)

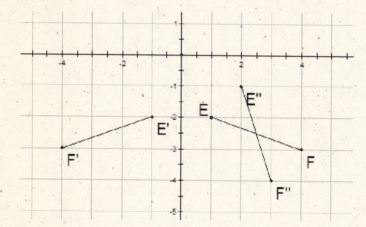

E"

E

E'

F'

F

F"

d) $E(1,-2) \rightarrow E'(-1,-2) \rightarrow E''(2,-1)$
   $F(4,-3) \rightarrow F'(-4,-3) \rightarrow F''(3,-4)$

e) $(a,b) \rightarrow (-b,-a)$

23.   a) The image of A is C under a 120^0 clockwise rotation about G.

        b) The image of A is F under a 60^0 counterclockwise rotation about G.

        c) The image of A is E under a 240^0 clockwise rotation about G.

        d) The image of A is E under a reflection about $\overline{CF}$.

25.   a)

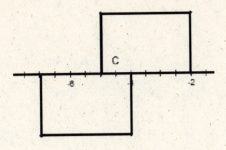

b)

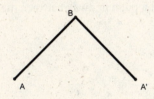

27.   a)

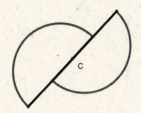

b)

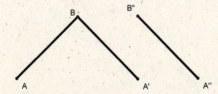

c) A reflection about *m*.

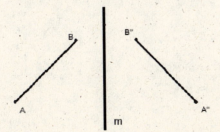

29.	a) The reflection is:

M

A

T

T

b) The reflection is:

T

I

M

31.	a) Suggests a rotation.

b) Suggests a reflection.

c) Suggests a translation.

33.	a) These are congruent and the transformation was a translation.

b) These are congruent and the transformation was 180^0 rotation about a point midway between the

bottoms of each F.

c) These shapes are not congruent.

35.	a) O is the center and there is an 180^0 rotation.

b) There are two lines of reflection.

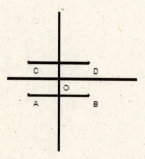

37.	If P is not on $\overleftrightarrow{AB}$, then $\overleftrightarrow{PP'} \| \overleftrightarrow{AB}$.

39.   a) (1) No   (2) No

        b) Only the first pair is congruent.

41.

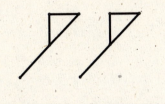

43.   a) 180^0 rotation.

        b) Reflection.

45.   The boy is photographed in one mirror while holding another.

47.   Under a translation in a plane, no points remain fixed.

49.   There is exactly one image for each rigid motion that is performed.

51.   a), b), c)

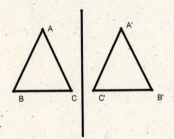

53.   If congruent is a reference to area then a possible solution is

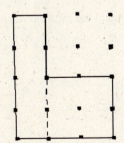

Lesson Exercises 9.2

LE1   Opener

         a) 180^0 rotation.

         b) $\angle ETO$

         c) A half turn around T appears to map $\angle ATO$ onto $\angle ETC$

LE2   Reasoning

         a) 180^0 rotation about A maps $\angle 1$ to $\angle 3$.

         b) $\angle 3$

         c) $\angle 1 \cong \angle 3 \cong \angle 5 \cong \angle 7$ and $\angle 2 \cong \angle 4 \cong \angle 6 \cong \angle 8$

LE3   Reasoning

         a) Alternate interior angles are congruent.

         b) Corresponding angles are congruent.

LE4   Concept

         $\angle 4$ and $\angle 1$, $\angle 4$ and $\angle 3$, $\angle 4$ and $\angle 5$, $\angle 4$ and $\angle 7$

LE5   Concept

         $\angle A \cong \angle D, \angle B \cong \angle E, \angle C \cong \angle F, \overline{AB} \cong \overline{DE}, \overline{AC} \cong \overline{DF}, \overline{BC} \cong \overline{EF}$

LE6   Concept

         a) $\angle A$

         b) $\overline{DA}$

LE7   Skill

Beetles: A reflection through a line that splits any beetle in half down the middle or a translation that moves

one white beetle left eye to the left eye of the white beetles above it.

Dogs: A translation up or to the right that moves the right ear of one white dog to the right ear of another

white dog.

LE8   Skill

a)

b)

c)

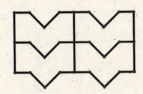

d)

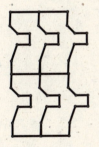

LE9   Concept

a)

b)

LE10   Concept

a)

b)

LE11   Summary

When a transversal cuts two parallel lines, alternate interior angles are congruent, alternate exterior angles are congruent, corresponding angles are congruent, and vertical angles are congruent. Rigid transformations can be used to determine if two figures are congruent. A tessellation is a design that can cover a plane without gaps.

Homework Exercises 9.2

1.   a) $\angle 6 \& \angle 8$, $\angle 1 \& \angle 3$

b) $\angle 6 \& \angle 4$, $\angle 3 \& \angle 5$

c) $\angle 6 \& \angle 2$, $\angle 3 \& \angle 7$

d) $\angle 6 \& \angle 5$

e) $\angle 1 = 110^0, \angle 2 = 70^0, \angle 4 = 70^0, \angle 5 = 110^0, \angle 6 = 70^0, \angle 7 = 110^0, \angle 8 = 70^0$

f) $\angle 7 \& \angle 8$ are supplementary

3.   a) 180^0 rotation about E.

b) $\overline{BA} \cong \overline{RT}$ and $\overline{BT} \cong \overline{RA}$

c) 180^0 rotation about E.

d) $\angle ART$

e) 180^0 rotation about E indicates that $\angle BTR \cong \angle RAB$

5.   a) $\angle ACB \cong \angle AGF, \angle ADE \cong \angle AHI, \angle ACD \cong \angle AGH, \angle ADC \cong \angle AHG$

b) $\angle BCG \cong \angle CGH, \angle DCG \cong \angle CGF, \angle EDH \cong \angle DHG, \angle CDH \cong \angle DHI$

7.   a) $\angle 2$ and $\angle 8$

b) Alternate exterior angles are congruent.

c) $\angle 1 \cong \angle 3$ Vertical angles

$\angle 3 \cong \angle 5$ Alternate interior angles

$\angle 5 \cong \angle 7$ Vertical angles

$\angle 1 \cong \angle 7$ Transitive Property

9.   a) If they have one side of equal length.

b) All rays are congruent.

11.   a) $\overline{ES} \cong \overline{DP}$

        b) $\angle ESU$

13.   a) Rotate 120^0 clockwise around a point where their "noses" meet.

        b) Reflect through the body line that passes between the eyes.

15.   a)

        b)

17.   For Student

19. $m\angle CBE = 52^0, m\angle EBD = 87^0, m\angle ABD = 41^0, m\angle BDA = 49^0$

21.   Opposite angles of a parallelogram are congruent.

23.   a) – d)

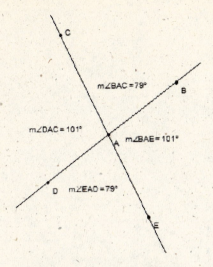

e) Vertical angles are congruent.

Lesson Exercises 9.3

LE1   Opener

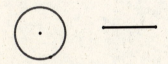

LE2   Skill

a) They are the same length.

b)

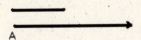

c) For student

d) For student

LE3   Connection

      a) Yes

      b) Yes

LE4   Skill

      a) – f) For student.

LE5   Concept

      (a) and (c) must be congruent by SSS.

LE6   Skill

      a) – f) For student.

LE7   Reasoning

      a) SSS triangle congruence

      b) $\angle A \cong \angle D$

      c) The triangles are congruent, so the corresponding angles are congruent.

LE8   Skill

      For student.

LE9   Reasoning

      Since $\overline{AC} \cong \overline{CD}$, $\overline{BC} \cong \overline{BC}$, and $\angle ACB \cong \angle DCB$, $\triangle ACB \cong \triangle DCB$ by SAS. If the triangles are congruent

      then $\overline{BD} \cong \overline{BA}$.

LE10   Skill

      a) – d) For Student

LE11   Reasoning

      a) $\overline{AB} \cong \overline{AC}, \overline{BD} \cong \overline{CD}, \overline{AD} \cong \overline{AD}$

      b) $\angle DBA \cong \angle DCA, \angle BAD \cong \angle CAD, \angle BDA \cong \angle CDA$

      c) Since $\triangle ABD \cong \triangle ACD, \angle BDA \cong \angle CAD$, which shows $\overline{AD}$ is the angle bisector of $\angle BAC$.

LE12   Skill

      For student.

LE13   Reasoning

a) By SSS, $\overline{AC} = \overline{BC}, \overline{AD} = \overline{BD}, \overline{CD} = \overline{CD}$

b) Because $\triangle BCD \cong \triangle ACD$

c) By SAS, $\overline{AC} = \overline{BC}, \angle ACM \cong \angle BCM, \overline{CM} = \overline{CM}$

d) Since $\triangle BCM \cong \triangle ACM$, $\overline{AM} = \overline{BM}$, so M is the midpoint of $\overline{AB}$

LE14   Reasoning

a) Shown in Figure 9 -54.

b) By SSS, $\overline{JA} = \overline{NA}, \overline{JM} = \overline{NM}, \overline{AM} = \overline{AM}$

c) Since $\triangle JAM \cong \triangle NAM$, $\angle MJA \cong \angle MNA$.

d) Deduction

LE15   Reasoning

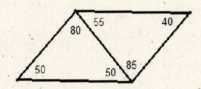

No, △*ABC* and △*ACD* are both isosceles triangles, but not necessarily congruent to one another. We do not

know what if any relationship exists between sides BC and CD. Here are two isosceles triangles that show

that when side DF changes so do the base angles.

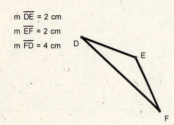

m $\overline{DE}$ = 2 cm
m $\overline{EF}$ = 2 cm
m $\overline{FD}$ = 4 cm

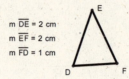

m $\overline{DE}$ = 2 cm
m $\overline{EF}$ = 2 cm
m $\overline{FD}$ = 1 cm

LE17   Skill

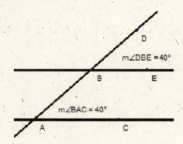

m∠DBE = 40°

m∠BAC = 40°

LE18   Reasoning

a) The perpendicular distance.

b) Construction of perpendicular for student.

LE19   Summary

Construction techniques allow for the duplication of geometric shapes which leads to investigations into

congruence properties.

Homework Exercise 9.3

1.   a)

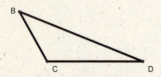

b) $\overline{MA} \cong \overline{BC}, \overline{MN} \cong \overline{BD}, \overline{AN} \cong \overline{CD}$
    $\angle AMN \cong \angle CBD, \angle MAN \cong \angle BCD, \angle ANM \cong \angle CDB$

SSM Chapter 9   338

3. a)

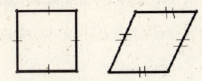

b) It indicates that a quadrilateral is not rigid.

5. a)

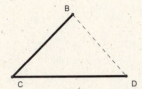

b) Notice the inclusion of side BD in the previous figure, using this, $\triangle MAY \cong \triangle BCD$ by SSS.

$\angle MAY \cong \angle BCD$ by corresponding parts of congruent triangles are congruent.

c) Deduction

7. a) SSS

b) The triangles are congruent so corresponding parts are congruent.

9. For student

11. a) No

b) Yes

c) No

d) Yes (SAS)

13.

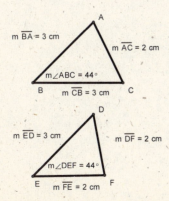

15.   a) Construction for student.

b), c) This construction works because of the SSS property. AC = AT and MC = MT because each pair of lengths were constructed with the same radius. AM = AM, Then $\triangle CAM \cong \triangle TAM$, by SSS. Since the triangles are congruent, $\angle MAC \cong \angle MAT$ which proves that $\overrightarrow{AM}$ bisects $\angle CAT$.

17.   a)

b)

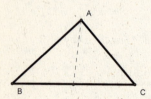

c) You should observe through paper folding the three medians intersect at a point, this point of intersection is called the centroid.

19.   a)

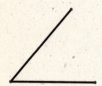

b) Paper folding for student.

21.   a) For student.

       b) For student.

23.

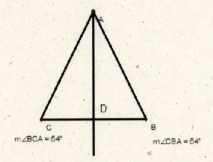

$\triangle ACD \cong \triangle ABD$ by AAS. $\overline{AB} \cong \overline{AC}$, corresponding parts of corresponding triangles are congruent.

25.   a) Each of the missing acute angles measures 45^0.

       b)

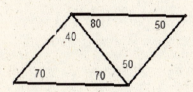

       c)

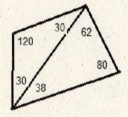

27.

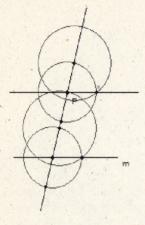

29.

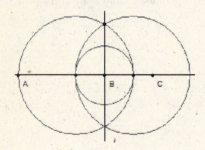

31.

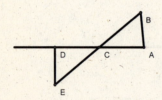

$\overline{AC} = \overline{CD}, \angle A \cong \angle D, \angle DCE \cong \angle ACB,$ Therefore $\triangle ACB \cong \triangle DCE$ by AAS. Then DE = AB.

33. $\triangle ABC \cong \triangle EFG$ SAS

AC = EG and $\angle BAC \cong \angle FEG$.

By subtraction $\angle DAC \cong \angle GEH$.

$\triangle ADC \cong \triangle EHG$ SAS

DC = HG and $\angle H \cong \angle D$

Since $\angle ACD \cong \angle EGH$ and $\angle ACB \cong \angle EGF$, we can add to obtain $\angle BCD \cong \angle FGH$

Therefore $ABCD \cong EFGH$

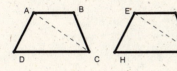

35.

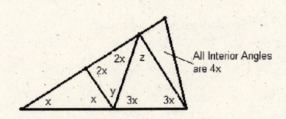

All Interior Angles are 4x

$$x + y + 3x = 180^0$$
$$4x + y = 180^0$$
$$6x + z = 180^0$$

The exterior angle of a triangle is equal to the sum of the two opposite interior angles. This yields

$$\angle A + \angle B + \angle C = 180^0$$
$$x + 4x + 4x = 180^0$$
$$9x = 180^0$$
$$x = 20^0$$

37.   Consider rhombus *ABCD*

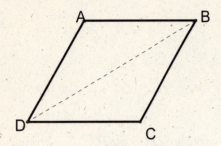

Prove ∠*DAB* ≅ ∠*DCB*

1.   *ABCD* is a rhombus

2.   *DA* ≅ *CB* Definition of a rhombus

3.   *DB* ≅ *DB* Reflexive Property

4.   △*DAB* ≅ △*DCB* SSS

5.   ∠*DAB* ≅ ∠*DCB* Corresponding parts of congruent triangles are congruent.

39.   a) Yes

b)

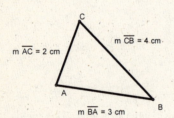

c) – f)

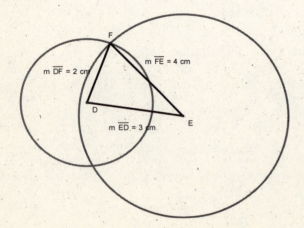

41.   a) – e)

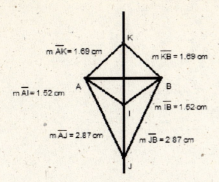

Points on a perpendicular bisector are equidistant from the endpoints of the segment that is bisected.

43.   For student.

Lesson Exercises 9.4

LE1   Opener

(a), (b), (d), (e), (g), (h) possess some type of symmetry.

LE2   Concept

For student

LE3   Connection

a) a, b, g, h, i, j

b)

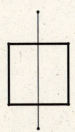

c) Fold along the reflecting line.

d) Place the Mira along the line of symmetry.

LE4   Reasoning

For student.

LE5   Concept

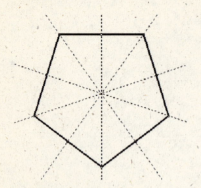

LE6   Concept

A reflection through the line does not leave the figure unchanged.

LE7   Skill

a)

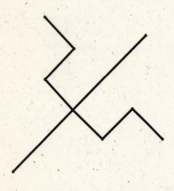

b)

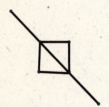

c)

LE8   Opener

a, d, e, and g

LE9   Concept

A regular pentagon has rotational symmetries of $72^o, 144^o, 216^o,$ and 288°.

LE10   Concept

a) (a) and (g).

b) (d) and (e).

c) (b), (h)

d) No this is not correct, a counter example could be drawn for the child such as:

LE11   Reasoning

LE12   Concept

Perpendicular to the plane of the wings and dividing the body in half.

LE13   Connection

Items could include a trash can, a chair, or a table.

LE14   Connection

a) An example could be a trash container.

b) The axis of rotation for a cylinder trash container would be a perpendicular line emanating from the

center of the circular base.

c) If the trash can is a cylinder and rotation will not change its appearance.

LE15   Summary

A plane figure has line symmetry if and only if it can be reflected across a line so that its image coincides with its original position. A plane figure has rotational symmetry if it can be rotated less than 360^0 so that it coincides with itself. Symmetry can be tested with tracing paper by first tracing the figure onto the paper then checking for reflections and rotations.

Homework Exercises 9.4

1.   Symmetry does enhance the beauty of shapes. Depending on your perspective one half may be pleasing to your eye.

3.   (a) has line symmetry.

5.   a), b), c)

d) The two lines of symmetry are perpendicular. The two lines of symmetry bisect the angles formed by the intersecting lines.

e) Induction

7. a) 4

    b) 2

    c) 2

    d) 0

9.

BIKE

11. This is not a line of symmetry for a rectangle. If the rectangle is reflected through the diagonal the image

    rectangle will not be in the same position as the original.

13. 0, 1, 2, 3, and 6 are possible solutions.

0

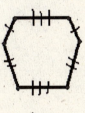

1

2

3

6

15. a) 3: 90°, 180°, 270°

    b) 1: 180°

    c) 1: 180°

    d) 1: 180°

17. (a) and (c) have rotational and reflection, (b) has reflection

19. a) Reflection symmetry, U, I, D

    b) Rotation symmetry, S, I

21. a) All 2's, 3's, 4's, 10's, J's, Q's, K's and the A, 3, 5, 6, 8, and 9 of diamonds.

    b) None

23.

25.   The number of rotation symmetries = The number of reflection symmetries – 1

27.   a)

        b)

29.   a) They are congruent.

        b) They are congruent.

31.   a) 4

        b) 1 Line of rotational symmetry for a right square pyramid, the axis through the center of the base to the

        apex.

33.   Right circular cone.

35.   3 and 1

37.   a) A tennis ball has rotational and reflectional symmetry if one ignores the ridges; a tennis racket has

        reflection symmetry; a tennis court has rotational symmetry; and the surface has reflection symmetry.

        b) The tennis ball will bounce well and can be hit at any point without it making a difference; one can strike

        the ball with either face of the of the tennis racket; the court on either side of the net is the same and the

        court is comparable on the left and right sides on each side of the net.

39.   a) Usually rotational

        b) Can be laid down a number of ways.

41.   a) TV screen, picture frame, table top, window, clock

        b) Table top, rug, window, chandelier, pillow

43.   a) Reflection

       b) Commutative

       c) Yes

45.   Fold the circle in half from two different positions to create two different diameters. The intersection of the

       two folds is the center.

47.   a)

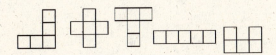

       b)

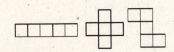

Lesson Exercises 9.5

LE1   Opener

       a) Shape

       b) Size

       c) 1: 2

       d) Shape

       e) Size

       f) 1: 3

LE2   Concept

       a) Not similar

       b) Similar 2: 3 ratio

LE3   Concept

       All squares are similar because corresponding angles are congruent and corresponding lengths are

       proportional.

LE4   Skill

a) $\triangle ABC \sim \triangle DEF$

b) c) $\dfrac{4}{x} = \dfrac{3}{5}$ $\dfrac{6}{y} = \dfrac{3}{5}$

$3x = 20$ $3y = 30$

$x = 6\dfrac{2}{3}$ $y = 10$

LE5   Skill

$\dfrac{4}{5} = \dfrac{8}{x}$ $\dfrac{4}{5} = \dfrac{6}{y}$

$4x = 40$ $4y = 30$

$x = 10$ $y = 7\dfrac{1}{2}$

LE6   Skill

Convert units of measure to inches.

$\dfrac{60}{94} = \dfrac{t}{250}$

$94x = 15000$

$x \approx 13 \, \text{feet} \, 3\dfrac{1}{2} \, \text{inches}$

LE7   Skill

$1 \, \text{inch} \approx 120 \, \text{miles}$

$\dfrac{5}{8} \, \text{inch} \approx 75 \, \text{miles}$

$1\dfrac{5}{8} \, \text{inches} \approx 195 \, \text{miles}$

a), b), c) The shape is a trapezoid

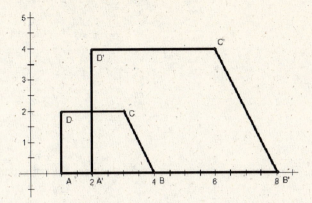

d) They are the same shape.

e) $A'B' = 2(AB); C'D' = 2(CD)$

f) They are equal.

g)

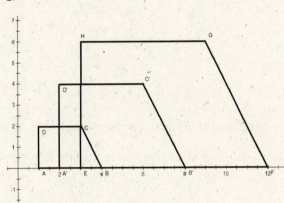

h) It is the same shape but larger.

i) $EF = 3(AB); GH = 3(CD)$

j) They are equal.

k) The size change multiplies the lengths of the original figure by the scale factor but does not change the

measures of the angles.

LE9 Reasoning

a), b)

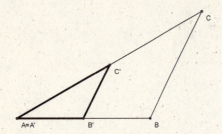

c) △*ABC* and △*A'B'C'* are the same shape. They have congruent corresponding angles. The corresponding sides of △*A'B'C'* are 0.5 times as long.

LE10 Reasoning

a) Corresponding angles are congruent. The sides are multiplied by the scale factor.

b) Similarity

c) Every line containing a point and its image intersects the center of dilation.

d) Same as c.

e) Same as c.

f) Induction.

LE11 Summary

Similar polygons have congruent corresponding angles and proportional corresponding lengths. A dilation is a transformation that may change the size but not the shape of a figure. The ratio of the dilated image to the original figure is called the scale factor.

Homework Exercises 9.5

1. Corresponding sides are not proportional, $\frac{3}{4} \neq \frac{4}{7} \neq \frac{5}{8}$

3. a) Two things are similar is frequently a reference to appearance.

b) If two shapes are similar in geometry the reference means items have congruent corresponding angles and proportional corresponding lengths.

5. a) $\angle T \cong \angle P$

b) $\frac{ER}{NA} = \frac{PE}{TN}$

7.	No, corresponding angles may not be congruent.

9.	You could first inquire as to the student's meaning of "same." If the reference is to proportions then we might agree with the child.

11.	$\dfrac{7}{20} = \dfrac{6}{x}$ 	$\dfrac{7}{20} = \dfrac{8}{y}$

	$7x = 120$	$7y = 160$

	$x = 17\dfrac{1}{7}$	$y = 22\dfrac{6}{7}$

13.	$\dfrac{4}{x} = \dfrac{10}{4}$

	$10x = 16$

	$x = 1.6$

15.	$\dfrac{3}{x} = \dfrac{6}{10}$ 	$\dfrac{6}{10} = \dfrac{7}{y}$

	$6x = 30$	$6y = 70$

	$x = 5$	$y = 11\dfrac{2}{3}$

17.	$\dfrac{52}{x} = \dfrac{70}{400}$

	$70x = 20800$

	$x = 297.14"$ or approximately $24.76'$

19.	a) 6.8 cm by 10.2 cm

	b) Yes

	c) The original area was $10 \times 15 = 150 \text{cm}^2$, the new area is $6.8 \times 10.2 = 69.36 \text{cm}^2$, so

	$\dfrac{69.36}{150} \times 100\% \approx 46.2\%$

21.	The dimensions should be:

	8 m would be represented by 2 cm

	10 m would be represented by 2.5 cm

	5 m would be represented by 1.25 cm

	4m would be represented by 1 cm

	3 m would be represented by 0.75 cm

23.

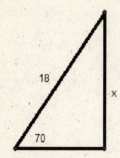

$$\sin 70^0 = \frac{x}{18}$$
$$x = 18\sin 70^0$$
$$x \approx 16.92'$$

25.

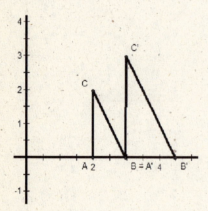

27.

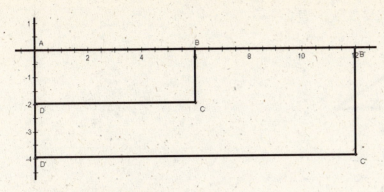

29.   The scale factor is 3 and the center is x.

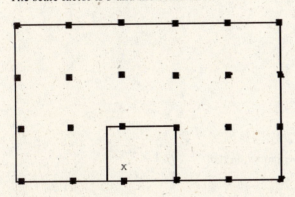

31.   a)

m∠ABC = 41°   m∠GHI = 41°

m∠CAB = 69°   m∠IGH = 69°

m∠ACB = 69°   m∠GIH = 69°

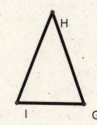

b) They are similar.

c) Corresponding sides are proportional.

d) Similar

e) Induction

33.   a)

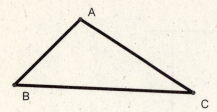

b)

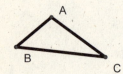

c) Yes

d) Continue to repeat (a) and (b).

e) Two triangles with proportional sides appear to be similar.

35.   1.10

37.   a) To dilate the pupil of the eye means to make the pupil larger.

b) In geometry dilation is a transformation that may change the size but not the shape of a figure.

39.   For student

41.   a), b), c), d), e), f)

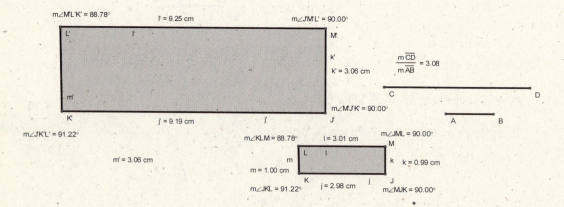

g) They are similar.

43.   For student

Chapter 9 Review Exercises

1.

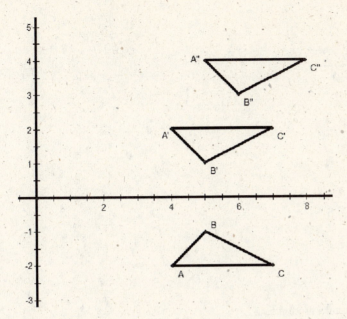

3.   a)

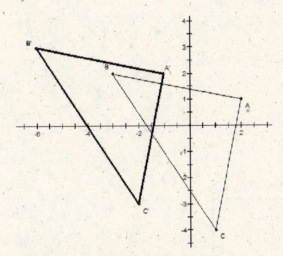

b) Translation left 3 units, up 1 unit

5.   Two figures are congruent if one can be mapped onto the other using a rotation, reflection, or translation.

7.   a) b) Note F' is the same point as E"

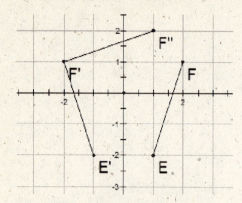

c) $E(1,-2) \rightarrow E'(-1,-2) \rightarrow E''(-2,1)$
   $F(2,1) \rightarrow F'(-2,1) \rightarrow F''(1,2)$

d) $(a,b) \rightarrow (b,a)$

9.   a) D

b) Could say $\angle DAB \cong \angle BCD$, 180^0 turn around E.

c) $\angle AEB \cong \angle CED$, 180^0 turn around E.

11.

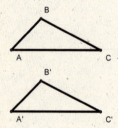

13. The angles of the hexagon are 120^0. Smaller angles of the triangle are 30^0; the remaining two angles

adjacent to the 30^0 angles are 90^0.

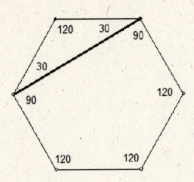

15. A square.

17.

0

1

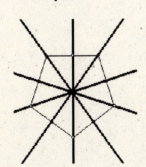

5

19. $\dfrac{RU}{RA} = \dfrac{8}{4}, \dfrac{UN}{AC} = \dfrac{8}{4}, UN = 10$

$\dfrac{10}{AC} = \dfrac{8}{4}$

$8AC = 40$

$AC = 5$

21.
$$\frac{L}{10} = \frac{L+6}{w}$$
$$Lw = 10(L+6)$$
$$w = \frac{10(L+6)}{L}$$
$$w = \frac{10L+60}{L}$$

Chapter 10

Lesson Exercises 10.1

LE1 Concept

a) About 3 or 4.

b) Their hands are of different sizes.

c) The person's hand is smaller.

LE2 Opener

a) There are 1,760 yards in a mile.

b) There are 1,000 meters in a kilometer.

LE3 Skill

a) Find a centimeter on a meter stick.

b) 100

c) Find a millimeter on a meter stick.

d) 10

e) 1,000

LE4 Skill

a) – d) Measurements for students to perform.

e) Your arm span should approximate your height.

LE5 Connection

a) Door knob.

b) Width of a doorway.

c) Width across a finger.

d) Thickness of a dime.

LE6 Concept

The best estimate should be choice (b).

LE7 Skill

For student.

LE8   Reasoning

For student

LE9   Connection

a) Yard

b) An inch is longer than a centimeter.

LE10   Skill

a) 0.5 m = 500 mm

b) 80 mm = 8 cm

c) A kilometer is 1,000 meters

LE11   Concept

80 mm = 8 cm. When you move from smaller to larger units, divide instead of multiply.

LE12   Concept

a) Same as

b) Less than

LE13   Connection

A raisin has a mass of about 1 gram.

LE14   Connection

Two loaves of bread would have a mass of about 1 kg.

LE15   Connection

A quart milk carton.

LE16   Skill

a) 80 g = 0.08 kg

b) N mL = $0.001N$ L

LE17   Connection

1 pint is 2 cups.

1 quart is 4 cups.

1 gallon is 16 cups.

LE18   Connection

Choice (a).

LE19   Skill

a) First count from 1:30 p.m. to 3:30 p.m. which is 2 full hours. Then add the time from 3:30 p.m. to 4:15 p.m. which is 45 minutes. The total elapsed time is 2 hours 45 minutes.

b) Start at 7:30 p.m. and count forward 1 hour to 8:30 p.m. Then count forward 40 minutes from 8:30 p.m. Some children would use 8 groups of 5 minutes, and some would go 30 minutes and another 10 minutes to reach 9:10 p.m.

LE20   Skill

a)   7 hr 30 min
    +1 hr 40 min
     8 hr 70 min

70 minutes is 1 hour and 10 minutes, so the movie ends at 9:10 p.m.

b) $\overset{3}{\cancel{4}}$ hr $\overset{75}{\cancel{15}}$ min

    −1 hr 30 min
     2 hr 45 min

LE21   Concept

a) About 25 cm.

b) About 253 mm.

c) Measuring to the nearest millimeter is more precise.

LE22   Concept

a) 3

b) 2

c) 3

LE23   Skill

     23.25
    +14.60
     37.85

    37.9 kg

LE 24   Skill

$$
\begin{array}{r}
9.2 \\
+14.8 \\
\hline
136.16
\end{array}
$$

140 ft^2

LE25   Summary

Every measurement requires a unit of measure. The metric system is based on a meter – roughly the

distance between the floor and a doorknob. Conversions in the metric system are easier.

Homework Exercises 10.1

1.   For student.

3.   a) A pencil.

        b) Approximately 2 paper clips.

5.   a) The child lined up the segment starting at 1 cm. The end of the segment was at 7 cm.

        b) The child should start the measurement at 0 cm.

7.   a) 3 feet = 36 inches

        b) 2 miles = 12,560 feet

        c) 5 feet = 15 yards

9.   a) Width across a finger.

        b) Distance from the floor to a doorknob.

11.   For student.

13.   a) 3 barley corns placed end to end.

        b) Length of an adult foot.

15. $2,800 \text{ miles} \times 1,760 \dfrac{\text{yards}}{\text{mile}} = 4,928,000 \text{ yards}$

17.   The number of units will increase.

19.   a) 0.02 m = 20 mm

        b) 16 cm = 160 mm

        c) 82 m = 0.082 km

21.	1 km = 1,000 m

	1,000m ÷ 50m = 20 lengths

23.	46 cm, 871 mm, 137 cm, 37 m, 3 km

25.	(d) 70 kg

27.	a) $4,000 ÷ 0.8 = 5,000$ paper clips

	b) Repeated measures.

29.	a) 7 pounds = 112 ounces

	b) 300 pounds = 0.15 ton

	c) 12 ounces = 0.75 pound

31.	3,240 milliliters

33.	a) 9.4 L = 9,400 mL

	b) 37 mg = 0.037 g

	c) 0.082 kg = 82 g

35.	(b) 250 mL

37.	$\dfrac{0.5 \text{ mg}}{2 \text{ mL}} = \dfrac{0.3 \text{ mg}}{x \text{ mL}}$

	$0.5x = 0.6 \text{ mL}$

	$x = 1.2 \text{ mL}$

39.	a) 3 gallons = 12 quarts

	b) 5 cups = 2.5 pints

41.	a) An ounce is 1 tablespoon of water.

	b) A quart is a milk carton.

43.	a) Meter stick

	b) 2 L of milk

	c) $16 \text{ cm} \times 22 \text{ cm}$ sheet of paper

	d) Hectogram hamburger

45.	a) g

	b) cm

	c) mm

47.   (a)

49.   (b)

51.   a) You could add one hour to 7:45 p.m. which would be 8:45 p.m. then count on 15 minutes to get to 9:00 p.m.

      b) Two hours added to 9:30 a.m. is 11:30 a.m. the additional 50 minutes is three quarters of an hour plus 5 minutes, so the meeting ending time is 12:20 p.m.

53.   a)   9 hr 30 min
          +2 hr 50 min
          ‾‾‾‾‾‾‾‾‾‾‾‾
          11 hr 80 min

      80 minutes is 1 hour and 20 minutes, so the meeting concluded at 12:20 p.m.

      b) $\overset{8}{\cancel{9}}$ hr $\overset{60}{\cancel{00}}$ min
          +7 hr 45 min
          ‾‾‾‾‾‾‾‾‾‾‾‾
          1 hr 15 min

55.   a) Approximate

      b) Exact

57.   a) – c) For student

      d) The measurement to the nearest eighth of an inch.

59.   a) 2

      b) 3

      c) 2

61.   80.3 cm

63.   0.76 cm^2

65. $300\dfrac{\text{km}}{\text{hr}} \times \dfrac{1\,\text{hr}}{60\,\text{min}} \times \dfrac{1\,\text{min}}{60\,\text{sec}} = 0.83\dfrac{\text{km}}{\text{sec}}$

      $0.83\dfrac{\text{km}}{\text{sec}} \times 2 \approx 0.166\,\text{km}$

      $\approx 0.17\,\text{km}$

67.   a) 3×10^{-6} sec

      b) 1×10^{-10} m

c) $\dfrac{1.6 \times 10^{-8}}{10^{-9}} = 1.6 \times 10$ nanoseconds or 16 nanoseconds

69.   For student

Lesson Exercises 10.2

LE1   Opener

    The site plan.

LE2   Reasoning

    a) The perimeter is 14 units.

    b) 18 is the largest perimeter. 12 is the smallest perimeter.

LE3   Concept

    $P = 2l + 2w$

LE4   Skill

    The floor plan.

LE5   Skill

    $C = 3.14 \times 26$
      $= 81.64$ inches

LE6   Concept

    a) 4

    b) 12

LE7   Reasoning

    For student

LE8   Concept

    a)

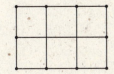

    b) $2m \times 3m = 6m^2$

LE9   Concept

No, there are not 10 mm^2 in a cm^2.

1 cm = 10

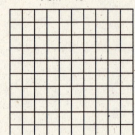

1 cm = 10

LE10   Concept

a) 2 rows, 4 columns

b) 8 sq cm

c) $l = 4$ cm, $w = 2$ cm

d) $A = lw$
    $A = 4(2)$
    $A = 8$cm^2

LE11   Concept

The site.

LE12   Connection

Package A: $2\frac{1}{2}' \times 6' = 15$ sq ft and $15' \times 3' = 45$ sq ft @ \$4.

Package B: $2' \times 5' = 10$ sq ft and $10' \times 4' = 40$ sq ft @ \$3.25.

Package A is 9 cents a square foot and package B is 8 cents per square foot. Package B is the better buy.

LE13   Concept

Yes

LE14   Reasoning

a)

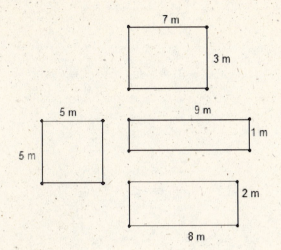

b) No

c) A longer, thinner shape.

LE15   Reasoning

a)

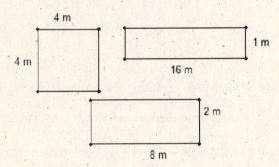

b) No

c) Square

LE16   Reasoning

a) False. The counterexample could be:

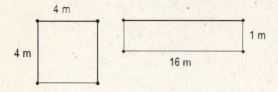

b) True

c) False. The counterexample could be:

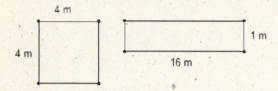

d) False. The counterexample could be:

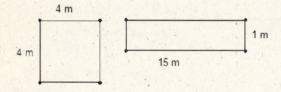

LE17   Summary

The perimeter of a simple, closed plane figure is the length of its boundary. The perimeter of a circle is referred to as the circumference. $C = \pi d$ or $C = 2\pi r$. Area is the measure of a closed two dimensional region. Area is expressed in square units. The area of a rectangle that has length l and width w is $A = lw$.

Homework Exercises 10.2

1.   a) The second rectangle.

b) Height

c) The first rectangle.

d) The area

3.   8.6 cm

5.   a) 12 units

b) 16 units is the largest perimeter and 12 units is the smallest perimeter.

7.   a) The second perimeter is twice as long as the first perimeter.

        b) For student

        c) When the length of each side of a square is doubled, the perimeter is doubled.

        d) Induction.

9. $2r = d$

11. $25,000 = \pi d$

$$d = \frac{25,000}{\pi}$$

$$d \approx 7,958 \text{ miles}$$

13.   a) $C = 2\pi r$

$$C = 2\pi(9.3 \times 10^7)$$

$$C \approx 5.84 \times 10^8 \text{ miles}$$

        b) $\left(5.84 \times 10^8 \div 365\right) \div 24 \approx 6.7 \times 10^4 \text{ miles}$

        c) Gravity

15.   a) $P = 2\pi r + 200$

$$P = 2\pi(20) + 200$$

$$P = 40\pi + 200$$

$$P \approx 325.68 \text{ m}$$

        b) 6.284 m

17.   3 hexagons

19.   Breaking the problem into pieces yields the following areas:

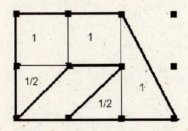

    The total area of the shaded region is 4 square units.

21.   a) – b) For student

23.   The 3 by 3 would yield more space for planting.

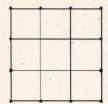

25.   Method 1: $A = (10\text{m} \times 50\text{m}) + (20\text{m} \times 50\text{m})$
        $A = (500 + 1,000)\text{m}^2$
        $A = 1,500\text{m}^2$

        Method 2: $A = 30\text{m} \times 50\text{m}$
        $A = 1,500\text{m}^2$

27.   For student

29. $0.2 \times 0.3 = 0.06$

31.   9 sq. ft. = 1 sq. yd.

33.   Not drawn to scale, each square is 1 decimeter by 1 decimeter

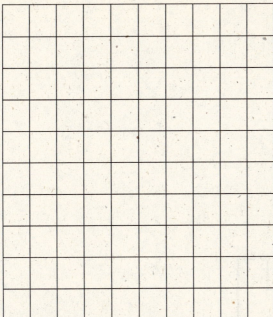

35.   a) $1\,\text{cm}^2 = 100\,\text{mm}^2$

        b) $1\,\text{cm}^2 = 0.0001\,\text{m}^2$

37.   a) $25\,\text{cm} \times 25\,\text{cm} = 625\,\text{cm}^2$

        $2\,\text{m} \times 4\,\text{m} = 8\,\text{m}^2$

        $8\,\text{m}^2 = 80,000\,\text{cm}^2$

        $80,000\,\text{cm}^2 \div 625\,\text{cm}^2 = 128\,\text{tiles}$

39.   Package A: $10\,\text{ft}^2 \times 6 = 60\,\text{ft}^2$, for \$5.25 $\Rightarrow 1\,\text{ft}^2$ for 9 cents

        Package B: $24\,\text{ft}^2 \times 4 = 96\,\text{ft}^2$, for \$7 $\Rightarrow 1\,\text{ft}^2$ for 7 cents

        Package B is the better buy.

41.

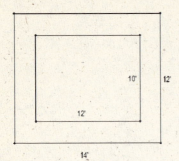

$$A = (12' \times 14') - (10' \times 12')$$
$$A = 48\,\text{ft}^2$$

43. a) Method 1:

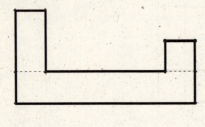

Method 2:

b) $A = (10' \times 18') + (22' \times 10') + (24' \times 10')$
  $A = 640\,\text{ft}^2$

  $C = 640\,\text{ft}^2 \times \$220 / \text{ft}^2$
  $C = \$140,800$

45. One answer is a 3 cm by 4 cm rectangle. Form other solutions by moving a square from one of the corners

to various other positions where it shares one side with another border square.

47.

49. The child is in error, it may happen that as the perimeter increases the area increases, but not always. A possible counter example:

Area = 100 square feet

Perimeter = 40 feet

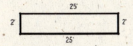

Area = 50 square feet

Perimeter = 54 feet

51. a) The smallest perimeter is associated with a square.

The perimeter is 8 meters.

b) The smallest perimeter is associated with a square.

The perimeter is 12 meters.

c) $4\sqrt{N}$ m

53. $2xy$

55.   a) 4, 6, 8

    b)

    R4 ⬚⬚⬚⬚

    Perimeter = 10

    c) $2N + 2$

    d) 22

57.   Each rectangle could have L = 42.5 feet and W = 60 feet.

59.

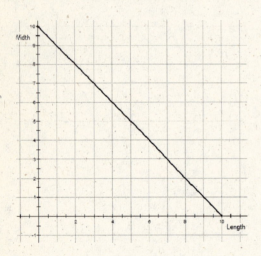

61.   a) $(a+2)^2 = a^2 + 2a + 4$

    b) $(a-b)^2 = a^2 - 2ab + b^2$

63.   The 500 miles is $\dfrac{7.5}{360}$ of the circle. The circumference would be $500 \cdot \dfrac{7.5}{360} = 24,000$ miles.

65.   a) $C3 = A3 * B3$

    b)

Rectangles	P=16	
Length	Width	Area
4	4	16
5	3	15
6	2	12
7	1	7

    c) The closer in size the length and width are, the greater the area.

67.   For student

SSM Chapter 10   378

Lesson Exercises 10.3

LE1   Opener

Area of a parallelogram: $A = bh$

Area of a triangle: $A = \frac{1}{2}bh$

Area of a trapezoid: $A = \frac{1}{2}h(b_1 + b_2)$

LE2   Skill

a) – b) For student

c) 24 cm

d) 24 cm

e) The rectangle has a larger area.

LE3   Reasoning

a) For student

b) 7cm by 4cm

c) 28 cm^2

d) 28 cm^2

LE4   Reasoning

a) Reflect the right triangle about h, then translate the right triangle to the left such that side h matches side AD to form a parallelogram.

b) $A = bh$

LE5   Skill

a) 50 square units

b) Deduction

LE6   Reasoning

a) – b) For student

c) bh

d) $\frac{1}{2}bh$

LE7   Skill

This is incorrect. The child needs to understand that the height is perpendicular to the base. Show that the height is 8 and the base is 7 so the area is 28 square units.

LE8   Reasoning

a) – b) For student

c) $h(b_1 + b_2)$

d) $A = \dfrac{1}{2} h(b_1 + b_2)$

LE9   Skill

a) About $A = \dfrac{1}{2}(110)(350 + 439)$

$A = 43,395$ square miles

b) Greater

LE10   Reasoning

a) A parallelogram

b) The base of the new figure is about $\dfrac{1}{2}$ the circumference of the original circle, and the height of the figure is about the same as the radius of the original circle.

c) $\dfrac{1}{2} Cr$

LE11   Skill

a) The amount of the large pizza is greater than the amount of the two smaller pizzas.

b) Area (large pizza): $A = \pi(8^2)$

$A = 64\pi$

Area (small pizza): $A = \pi(4^2)$

$A = 16\pi$

Two small pizzas have an area of: $A = 16\pi + 16\pi$

$A = 32\pi$

One 16 inch pizza has twice as much pizza as two 8 inch pizzas.

LE12   Reasoning

Area (semicircle): $A = \frac{1}{2}\pi r^2$

$$= \frac{1}{2}\pi(5^2)$$

$$= \frac{25\pi}{2} \text{ square units}$$

Area (triangle): $A = \frac{1}{2}bh$

$$= \frac{1}{2}(6)(8)$$

$$= 24 \text{ square units}$$

Shaded Area: $A = \frac{25\pi}{2} \text{ square units} - 24 \text{ square units}$

$A = (12.5\pi - 24) \text{ square units}$

LE13   Summary

The area of a parallelogram: $A = bh$

The area of a triangle: $A = \frac{1}{2}bh$

The area of a trapezoid: $A = \frac{1}{2}h(b_1 + b_2)$

The area of a circle: $A = \pi r^2$

Homework Exercises 10.3

1.   a) 6 sq. units

b) 4 ½ sq. units

3.   (b) Figure 2

5.   a) A = 50 sq. in

P = 32 in

b) A = 6 sq. m

P = 12 m

7.   A parallelogram can be rearranged to form a rectangle. Cut off the right triangle, reflect it about the height, then translate it to the opposite side.

9.   Put together two of the triangles as shown to form a parallelogram.

11.   a) $A = \frac{1}{2}(30)(24)$

$A = 360 \, \text{in}^2$

b) $A = \frac{1}{2}(6)(4)$

$A = 12 \, \text{units}^2$

c) $C = \frac{\$0.79}{100 \, \text{in}^2} \times 360 \, \text{in}^2$

$C = \$2.84$

13.

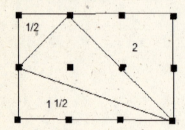

Area of the triangle can be found be subtracting the shown areas from the area of the constructed rectangle.

A = 6 − (2 + 1 ½ + ½)

A = 2 square units

15.   Put together two of the trapezoids shown to form a parallelogram.

$A_{\text{trapezoid}} = \frac{1}{2} A_{\text{parallelogram}} = \frac{1}{2} h(b_1 + b_2)$

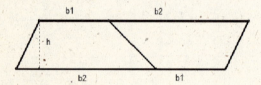

17.   About 270 mm^2

19.   B ($28.57/m^2)

21.   Squares and rectangles

23.   a) $A = \frac{1}{2}Cr = \frac{1}{2}(2\pi r)r = \pi r^2$

b) Deduction

25.   a) 20.25π units^2

        b) 63.585 units^2

        c) $\left(\dfrac{8}{9}d\right)^2$ or $\left(\dfrac{16}{9}r\right)^2$ or $\dfrac{256}{81}r^2$

27.   a) $A_{14"} = \pi(7)^2$
            $A_{14"} = 49\pi\,\text{in}^2$

            $A_{10"} = \pi(5)^2$
            $A_{10"} = 25\pi\,\text{in}^2$

            $\dfrac{49\pi}{25\pi} = 1.96$ larger

29.   a) 2

        b) 4

31.   rectangle, triangle, parallelogram, circle

33.   a) $P = \dfrac{5}{6}(2\pi 4) + 4 + 4$

          $P = \left(\dfrac{20\pi}{3} + 8\right)$ units

            $A = \dfrac{5}{6}\pi(4)^2$
        b)
            $A = \dfrac{40\pi}{3}$ units^2

35. $2\pi r = 20\pi$ $A = \pi(10)^2$
          $r = 10\,\text{ft}$ $= 100\pi\,\text{ft}^2$

37. $A_{\text{path+garden}} = \pi(4)^2$
                 $= 16\pi\,\text{m}^2$

        $A_{\text{garden}} = \pi(3)^2$
              $= 9\pi\,\text{m}^2$

        $A_{\text{path}} = 16\pi\,\text{m}^2 - 9\pi\,\text{m}^2$
            $= 7\pi\,\text{m}^2$

39. $A_{shaded} = \left(\dfrac{1}{2} \pi(3)^2 - \dfrac{1}{2} \pi(1.5)^2 \right) \text{units}^2$

$\qquad = 2.25\pi \text{ units}^2$

41. $A = \pi (1)^2$

$A = \pi \text{ units}^2$

$A_{figure} = \pi \text{ units}^2 + \pi \text{ units}^2 - \dfrac{\pi}{4} \text{units}^2$

$A_{figure} = \dfrac{7}{4} \pi \text{ units}^2$

43. $9\pi - 18 \text{ units}^2$

$A_{sector} = \dfrac{1}{4} \pi(6)^2 \text{ units}^2$

$A_{sector} = 9\pi \text{ units}^2$

$A_{triangle} = \dfrac{1}{2}(6)(6) \text{ units}^2$

$A_{triangle} = 18 \text{ units}^2$

45. (c)

47. a) – b) For student

c) $A_{EFGH} = \dfrac{1}{2} A_{ABCD}$

49. $A = \pi r^2 \text{ units}^2$

$A = \pi \left(\dfrac{C}{2\pi} \right)^2 \text{ units}^2$

$A = \dfrac{C^2}{4\pi} \text{ units}^2$

51. For student

Lesson Exercises 10.4

LE1   Reasoning

a) 1 sq. unit, 1 sq. unit, 2 sq. units

b)

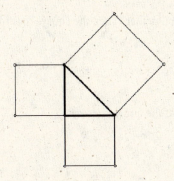

c) 4 sq. units, 4 sq. units, 8 sq. units

d) 1 sq. unit, 4 sq. units, 5 sq. units

e) $c^2 = a^2 + b^2$

LE2   Reasoning

a) Each side has length $a + b$, and it has four right angles.

b) 90^0

c) 90^0

d) It has four 90^0 angles and four congruent sides.

e) $(a+b)^2$

f) $4\left(\dfrac{1}{2}ab\right) + c^2$

g) $(a+b)^2 = 4\left(\dfrac{1}{2}ab\right) + c^2$

$\quad a^2 + 2ab + b^2 = 2ab + c^2$

$\quad\quad\quad a^2 + b^2 = c^2$

LE3   Connection

a) $N = \sqrt{2}$

b) Irrational numbers

LE4   Skill

a) $c^2 = 14^2 + 18^2$

$c^2 = 520$

$c = \sqrt{520}$

$c \approx 22.8\,\text{inches}$

b) This is not correct. 14^2 and 18^2 must first be computed and added together. Then the square root of this sum is taken.

LE5   Reasoning

Let c represent the diagonal of the rectangle, it is also the diameter of the circle.

$c^2 = 12^2 + 16^2$

$c^2 = 400$

$c = \sqrt{400}$

$c = 20\,\text{units}$

$c = 2r$

$20 = 2r$

$10\,\text{units} = r$

$A_{circle} = \pi(10)^2$

$\qquad = 100\pi\,\text{units}^2$

$A_{shade} = (100\pi - 192)\,\text{units}^2$

LE6   Skill

If a, b, and c are the lengths of the sides of a triangle and $c^2 = a^2 + b^2$, then the triangle is a right triangle.

LE7   Skill

a) $10^2 \stackrel{?}{=} 6^2 + 8^2$ This is a right triangle.

$100 = 100$

b) $9^2 \stackrel{?}{=} 6^2 + 7^2$ This is not a right triangle.

$81 \neq 85$

c) Deduction

LE8   Concept

For student

LE9 Reasoning

a) AB

b) It is longer than AB.

c) $AC + CB > AB$

LE10 Skill

a) These dimensions could form a triangle since the triangle inequality is satisfied.

b) These dimensions will not work since the sum of 8 and 9 is less than 20.

LE11 Reasoning

a) $4 < c < 7$

b) Right triangle for $c = 5$

Obtuse triangle for $5 < c < 7$

Acute triangle for $4 < c < 5$

LE12 Reasoning

a) $a + b > c$

b) $a^2 + b^2 = c^2$ For a right triangle

$a^2 + b^2 > c^2$ For an acute triangle

$a^2 + b^2 < c^2$ For an obtuse triangle

LE13 Reasoning

a) Acute triangle $6^2 + 7^2 > 9^2$

b) Not a triangle $3 + 7 < 12$

c) Obtuse triangle $4^2 + 5^2 > 8^2$

LE14 Summary

If a right triangle has legs of lengths a and b and a hypotenuse of length c, then $a^2 + b^2 = c^2$. The sum of the measures of any two sides of a triangle is greater than the measure of the third side. Problems involving length can be solved using the Pythagorean theorem.

Homework Exercises 10.4

1. a) Right triangle

    b) a and b are sides, c is the hypotenuse.

3. a) $A_1 = \frac{1}{2}ba$, $A_2 = \frac{1}{2}ba$, $A_3 = \frac{1}{2}c^2$

    b) $A_{trapezoid} = \frac{1}{2}(a+b)(a+b)$

    $\qquad\qquad = \frac{1}{2}(a+b)^2$

    c) $\frac{1}{2}ab + \frac{1}{2}ab + \frac{1}{2}c^2 = \frac{1}{2}(a+b)(a+b)$

    $\qquad\qquad 2ab + c^2 = \left(a^2 + 2ab + b^2\right)$

    $\qquad\qquad\qquad c^2 = a^2 + b^2$

    d) Deduction

5.

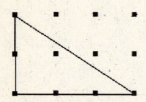

7. $P = \left(4\sqrt{2} + 6\right)$ units

9. $a^2 + 0.8^2 = 4^2$

    $\qquad a^2 = 15.36$

    $\qquad a = 3.92\,\text{m}$

11. $3^2 + 8^2 = c^2$

    $\qquad 73 = c^2$

    $\sqrt{73}\text{ft} = c$

    A 9 foot glass will not fit through the doorway.

13. $4^2 + 4^2 = c^2$ to the northeast.

    $\qquad 32 = c^2$

    $4\sqrt{2}\text{ mph} = c$

15.

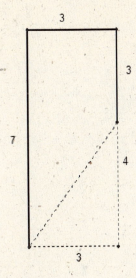

$$3^2 + 4^2 = d^2$$
$$5 \text{ miles} = d$$

17.

$$4^2 + h^2 = 6^2$$
$$h^2 = 20$$
$$h = 2\sqrt{5}$$

$$A = \frac{1}{2}(8)(2\sqrt{5})$$
$$A = 8\sqrt{5}$$

$$8\sqrt{5} + 8\sqrt{5} = 16\sqrt{5} \text{ units}$$

19.

$$5^2 + 8^2 = c^2$$
$$89 = c^2$$
$$\sqrt{89} = c$$

$$A_{bigcircle} = \pi\left(\sqrt{89}\right)^2$$
$$= 89\pi$$

$$A_{smallcircle} = 25\pi$$

$$A_{shade} = 89\pi - 25\pi$$
$$= 64\pi \text{ units}^2$$

21. $P = 15 + 10 + 21 + 8$
   $= 54$ m

   $A = \frac{1}{2}(8)(15 + 21)$
   $= 144$ m^2

23.   a) 10

   b) Finding the length of the hypotenuse.

25.   a) Yes

   b) No

27.   a) Yes

   b) 6, 8, 10; 5, 12, 13; 9, 12, 15

   c) If $a^2 + b^2 = c^2$ then $k\left(a^2 + b^2\right) = kc^2$ or $ka^2 + kb^2 = kc^2$. So ka^2, kb^2, kc^2 is a Pythagorean triple.

29.   The airports could be 50 miles apart.

31.   a) Obtuse

   b) Right

   c) Not a triangle

   d) Obtuse

33.   Row 1: v = 5; Row 2: u = 64; Row 3: c = 6649, v = 32; Row 4: b = 12,709, u = 125;

   Row 5: a = 72, u = 9, v = 4

35.   a) $x = 3, y = 4, z = 5; x = 6, y = 8, z = 10; x = 9, y = 12, z = 15$

   b) No

37. $16^2 + 12^2 = c^2$
      $400 = c^2$
       $20 = c$

   $A = \frac{1}{2}\pi(10)^2$
      $= 50\pi$ units^2

39. $$4^2 + 8^2 = c^2$$
$$80 = c^2$$
$$4\sqrt{5} = c$$

This is the diagonal of one of the faces of the closet

The diagonal of the closet is:

$$6^2 + \left(4\sqrt{5}\right)^2 = c_1^2$$
$$116 = c_1^2$$
$$\sqrt{116} = c_1$$
$$10.7 \approx c_1$$

The pole will fit.

41.   a) $a^2 + b^2 = c^2$

b) They both equal $a^2 + b^2$

c) $\triangle GFE$, SSS property

d) $\angle D$

43.   a) – g) For student

h) The sum of the areas off the sides is equal to the sum of the area off the hypotenuse.

i) For student

45.   For student

47.   For student

Lesson Exercises 10.5

LE1   Opener

If you want to gift wrap a package you will need to know the surface area of the package.

LE2   Skill

a)

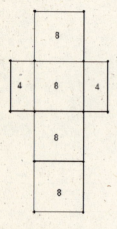

b) A = 40 sq. units

c) Discuss whether she has counted all the surfaces.

LE3   Skill

a)

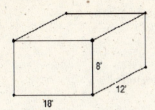

b) Area of ceiling $=$ 216 sq. ft.

c) Area of wall $=$ 96 sq. ft.

4 walls + ceiling $=$ 696 sq. ft.

Area of window $=$ 15 sq. ft.

Area of door $=$ 28 sq. ft.

Area to be painted $= 696 - (30 + 28) = 638$ sq. ft.

d) $638 \div 400 = 1.595$, so 2 gallons of paint must be purchased.

LE4   Skill

First find the hypotenuse of the 5, 12 right triangle.

$$5^2 + 12^2 = c^2$$
$$169 = c^2$$
$$13 = c$$

$$A = \frac{1}{2}(5)(12) + \frac{1}{2}(5)(12) + (12)(20) + (5)(20) + (13)(20)$$
$$= 660 \text{ m}^2$$

LE5   Reasoning

a) 3

b) Circular

c) Rectangular

LE6 Reasoning

a) $2\pi r$ by h

b) $A_{circle} = \pi r^2$

$A_{rectangle} = 2\pi rh$

The surface area of the cylinder is $A_{circle} + A_{circle} + A_{rectangle} = \left(2\pi r^2 + 2\pi rh\right)$ units^2

LE7 Skill

a)

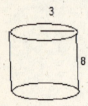

b) $A_{bases} = 9\pi + 9\pi$

$= 18\pi$ cm^2

c) $A_{lateral\ surface} = 6\pi(8)$

$= 48\pi$ cm^2

d) $A_{cylinder} = 18\pi$ cm$^2 + 48\pi$ cm^2

$= 66\pi$ cm^2

LE8 Reasoning

a) Find the perimeter of the base and multiply by the height.

b) $A_{surface} = ph + 2B$

LE9 Reasoning

a) $A = (5)(5) + 4\left(\dfrac{1}{2}(5)(4)\right)$

$= 65$ ft^2

b) $A = s^2 + 4\left(\dfrac{1}{2}sl\right)$ units^2

LE10   Reasoning

a) $A = B + n\left(\dfrac{1}{2}sl\right)$ units^2

b) $P = ns$ units

c) $A = B + \left(\dfrac{1}{2}Pl\right)$ units^2

LE11   Skill

a) $4^2 + b^2 = 5^2$

$\,9 = b^2$

$\,3 = b$

b) $A = 8^2 + \left(\dfrac{1}{2}(32)(3)\right)$ cm^2

$ = 112$ cm^2

LE12   Reasoning

a) $B = \pi r^2$ units^2 $\quad P = \dfrac{1}{2}(2\pi r)l$ units^2

b) $A = \pi r^2 + \dfrac{1}{2}(2\pi r)l$ units^2

LE13   Skill

$(3\pi)^2 + b^2 = 13^2$

$\,b^2 = 169 - 9\pi^2$

$\,b \approx 80$

$A = \dfrac{1}{2}(6\pi)(13)$ cm^2

$ = 39\pi$ cm^2

LE14   Summary

The surface area for a prism or cylinder, $A = ph + 2B$ where p is the perimeter, h is the height, and B is the

area of the base. The surface area for a right regular pyramid or cone is $A = B + \dfrac{1}{2}pl$.

Homework Exercises 10.5

1.   a)

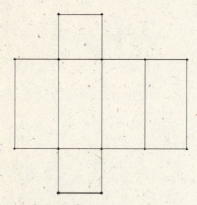

    b) 32 sq units

3.   a) $1 \times 1 \times 20, 1 \times 2 \times 10, 1 \times 4 \times 5, 2 \times 2 \times 5$

    b) $2 \times 2 \times 5$

5. $A = 30(20) + 30(20) + 15(20) + 15(20) + 30(15) + 30(15)$
        $= 2,700 \text{ cm}^2$

7. $A = 2lw + 2lh + 2wh \text{ m}^2$
    $246 = 2(8)(3) + 2(8)w + 2(3)w$
    $198 = 22w$
    $9 \text{ m} = w$

9. $A_{4\,walls} = 2(16)(8) + 2(8)(10) \text{ ft}^2$
           $= 416 \text{ ft}^2$

    $A_{door} = 28 \text{ ft}^2$

    $A_{window} = 18 \text{ ft}^2$

    $A_{ceiling} = 160 \text{ ft}^2$

    $A_{paint} = \left(460 + 160 - (28 + 18) \right) \text{ ft}^2$
           $= 530 \text{ ft}^2$

Since a gallon of paint covers 450 square feet, you need 2 gallons.

11.   a) $A = (4,050 + 8,100)$ cm^2

          $= 12,150$ cm^2

      b) 12,150 cm^2 = 1.215 m^2

      1.215 m$^2 \times \$20/m^2 = \$24.30$

13. $A = 50 + 70 + 10\sqrt{24} + \dfrac{1}{2}\sqrt{24}(5) + \dfrac{1}{2}\sqrt{24}(5)$ units^2

          $= 120 + 15\sqrt{24}$ units^2

15. $A = 36\pi + 36\pi + 12\pi(15)$ in^2

          $= 252\pi$ in^2

17.   Find the area of the base, πr^2. Since there are 2 bases, the area of both would be $2\pi r^2$. Next find the

      lateral area, which is the area of the rectangle, $2\pi rh$. The total surface area is $2\pi r^2 + 2\pi rh$.

19.   a) 916 sq ft

      b) $916 \div 400 = 2.29$ gal, need 3 gallons

      c) 3

21.   a) $A = \dfrac{5sl}{2}$

      b) $5s$

23.   a)

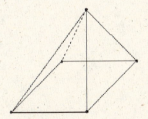

      b) Slant height: $3^2 = 2^2 + b^2$

                $5 = b^2$

                $\sqrt{5} = b$

      c) $A = 16 + 4\left(\dfrac{1}{2}\right)(4)(\sqrt{5})$

          $= 16 + 8\sqrt{5}$ m^2

25. $A = \pi r l$

        $= 7.5\pi$ in^2

27.   a) $F1 = 2 + 4x$ $F2 = 2 + 8x$

        b)

        $F3 = 2 + 12x$

        c) $F10 = 2 + 40x$

        d) $FN = 2 + 4Nx$

29.   Multiplies by 9.

        If the edge length is s: $S.A. = 6s^2$

        If the edge length is $3s$: $S.A. = 6(3s)^2$

        $ = 6(9)s^2$

        $ = 9\left(6s^2\right)$

31.   a) $A_{smallball} = 4\pi(2)^2$ in^2

           $\phantom{A_{smallball}} = 16\pi$ in^2

        $A_{largeball} = 4\pi(4)^2$ in^2

        $\phantom{A_{largeball}} = 64\pi$ in^2

        b) The larger ball has a surface area 4 times greater than the smaller ball.

        c) Cost is 4 times as great.

33.   For student

Lesson Exercises 10.6

LE1   Opener

        To know how many gallons of water a swimming pool holds, you would measure its volume.

LE2   Concept

        a) 12 cubic units

        b) 38 sq units

c) Volume would be the same, the surface area would be 32 sq units.

LE3   Reasoning

a) 12

b) 2

c) 24 cm^3

d) 4 cm, 3 cm, 2 cm

e) $4 \times 3 \times 2 = 24$ cm^3

LE4   Connection

a) Freezer A: 11.25 ft^3 for $310; $27.56 per cubic foot

Freezer B: 14 ft^3 for $400; $28.57 per cubic foot

Freezer A has the lower unit cost.

b) You can compare them either way, but it changes whether you would prefer a larger or smaller number.

LE5   Reasoning

a) Circular

b) πr^2

c) $V = \pi r^2 h$

LE6   Skill

a) $V = \pi (5)^2 h$
   $= 250\pi$ cm^3

b) $7^2 = 4^2 + x^2$
     $33 = x^2$
     $\sqrt{33} = x$

   $V = \dfrac{1}{2}(4)(\sqrt{33})(20)$
      $= 40\sqrt{33}$ in^3

LE7   Concept

a) They are equal.

b) They are equal

LE8   Reasoning

$$V = \frac{1}{3}\pi r^2 h$$

LE9   Skill

$$V = \frac{1}{3}(8 \bullet 8 \bullet 6)$$
$$= 128 \text{ units}^3$$

LE10   Reasoning

a) $V_{cone} = \frac{1}{3}\pi r^2 h$

$V_{cylinder} = \pi r^2 h$

b) $V = \frac{1}{2}\left(\frac{4}{3}\pi r^3\right)$

$= \frac{2}{3}\pi r^3$

c) $V = \frac{4}{3}\pi r^3$

LE11   Reasoning

a) $V \approx \frac{4}{3}\pi(1)^3 + \frac{1}{2}\left(\frac{4}{3}\right)\pi(1)^3$

$\approx 6.3 \text{ in}^3$

b) $6.3 \div 1.8 = 3.5$

LE12   Summary

Volume is the amount of space occupied by a three dimensional figure. The volume of a right rectangular

prism is $V = lwh$. The volume of any right prism or cylinder is $V = Bh$. The volume of a pyramid or cone

is $V = \frac{1}{3}Bh$.

Homework Exercises 10.6

1.   11 cubic units

3.   a) 30 cubic units

b) The child counts the number of squares that are visible.

5.   Freezer A: 18 ft^3 for \$350; \$19.44 per cubic foot

Freezer B: 31.5 ft^3 for \$480; \$15.24 per cubic foot

Freezer B has the lower unit cost.

7.   There are 10^6 cm^3 in a m^3.

9.   a) 1 cm^3

b) $1\text{mL} = 0.001\text{L}$
$$= 1\text{g}$$

$(4,000)0.001\text{L} = (4,000)1\text{g}$
$$4\text{L} = 4,000 \text{ g}$$

c) $70 \times 50 \times 40 = 140,000$ cm^3

$1\text{cm}^3 = 1$ mL

$(140,000)1\text{cm}^3 = (140,000)1$ mL
$$140,000\,\text{cm}^3 = 140,000 \text{ mL}$$
$$= 140 \text{ L}$$

d) $1\text{cm}^3 = 1$ g

$1\text{m}^3 = 1,000$ kg

e) $1\text{m}^3 = 1,000,000$ mL^3
$$= 1 \text{ kL}$$

11.   a) Surface area

b) Volume

c) Volume

13. $1^2 = 0.5^2 + x^2$

$$\frac{3}{4} = x^2$$

$$\frac{\sqrt{3}}{2} = x$$

$$V = \frac{1}{2}\left(\frac{\sqrt{3}}{2}\right)(0.5)(3)$$

$$= \frac{1.5\sqrt{3}}{2}$$

$$\approx 1.3 \text{ m}^3$$

15. $V = Bh$

$\quad = \pi(5.3)^2(17.5)$

$\quad \approx 1{,}544.3 \text{ cm}^3$

17. $400 = 36\pi h$

$\quad \dfrac{400}{36\pi} = h$

$\quad 3.54 \approx h$

19. $V = \pi R^2 h - \pi r^2 h$

$\quad = \pi(2.5)^2(4.5) - \pi(0.75)^2(4.5)$

$\quad = 28.125\pi - 2.53\pi$

$\quad \approx 25.59 \text{ in}^3$

21. a) $V = \pi r^2 h$

$\quad = \pi(3.1)^2(12)$

$\quad \approx 362 \text{ cm}^3$

b) $355 \text{ mL} = 355 \text{ cm}^3$

23. $V_{water} = \pi(2.5)^2(12)$

$\quad \approx 23{,}550 \text{ in}^3$

This is approximately 102 gallons.

Total weight: $200 + 102(8) = 1{,}016 \text{ lb}$

25. $3^2 + h^2 = 10^2$

$\quad h^2 = 91$

$\quad h = \sqrt{91}$

$V = Ah$

$\quad = 50(\sqrt{91})$

$\quad \approx 476.97 \text{ m}^3$

27. a) $V = \dfrac{1}{3}Bh$ applies to a cone or pyramid with base area B and height h.

b) $V = Bh$ applies to a cylinder or prism with base area B and height h.

29. a) $V = \dfrac{1}{3}Bh$

$ = \dfrac{1}{3}(1,960,000)(177)$

$ = 115,640,000 \text{ ft}^3$

b) $V = lwh$

$ = 25(30)(10)$

$ = 7,500 \text{ ft}^3$

$115,640,000 \div 7,500 \approx 15,419 \text{ Apartments}$

31. First find the area of the base, πr^2 and then substitute this into $V = Bh$.

33. $V = \dfrac{1}{3}\pi(1)^2(5.5)$

$ \approx 5.75 \text{ in}^3$

35. a) $V = 180$

b) Omitting π from the answer.

37. $V = \dfrac{4}{3}\pi r^3$

$ = \dfrac{4}{3}\pi\left(\dfrac{7926}{2}\right)^3$

$ \approx 2.607 \times 10^{11} \text{ mi}^3$

39. $A = \dfrac{1}{2}h(b_1 + b_2)$

$ = \dfrac{1}{2}(40)(1+3)$

$ = 80 \text{ m}^2$

$A = lw$

$ = (3)(10)$

$ = 30 \text{ m}^2$

Total Area: 110 m^2

$V = 110(20)$

$ = 2,200 \text{ m}^3$

$1 \text{ L} = 1,000 \text{ cm}^3$

$2,200 \text{ m}^3 = 2,200,000,000 \text{ cm}^3$

$\phantom{2,200 \text{ m}^3} = 2,200,000 \text{ L}$

41. $V_{can} = \pi r^2 l$

$\approx \pi(14.4)(19.3)$

≈ 873.1 cm^3

$V_{ball} = \dfrac{4}{3}\pi r^3$

$\approx \dfrac{4}{3}\pi(32.8)$

≈ 137.3 cm^3

3 tennis balls have an approximate volume of: 411 cubic cm.

$\dfrac{411}{873} \times 100\% \approx 47\%$

43. Draw radii to points of contact, $r = 30$ ft, 113,097 ft^3

45. a)

l	w	h
20	20	20
40	10	20
40	5	40

b)

length	width	height	surface area
20	20	20	2400
40	10	20	2800
40	5	40	4000

Being closer to a cube.

47. Lay it on its side.

49. a) Same material, less expensive.

b) $\dfrac{V}{\pi r^2} = h$

c) $A = 2\pi r^2 + 2\pi rh$

d) $A = 106.18$ cm^2

e) $r = 2.37$ cm, $h = 4.75$ cm

51. For student

Lesson Exercises 10.7

LE1   Opener

The second piece of land is 4 times as big and only costs three times as much. It is a better buy.

LE2   Skill

a) $30^2 + 40^2 = 50^2$

$900 + 1,600 = 2,500$

$60^2 + 80^2 = 100^2$

$3,600 + 6,400 = 10,000$

b) $A_1 = \frac{1}{2}(30)(40)$

$= 600 \text{ ft}^2$

$A_2 = \frac{1}{2}(60)(80)$

$= 2,400 \text{ ft}^2$

c) 4

d) Perimeter 1: 120 ft

Perimeter 2: 240 ft

e) 2

f) The second piece of land.

LE3   Skill

a) $\frac{2}{3}$

b) 24 cm^2 and 54 cm^2

c) $\frac{4}{9}$

d) 20 cm and 30 cm

e) $\frac{2}{3}$

LE4   Reasoning

a) $m^2 : n^2$

b) $m : n$

LE5   Reasoning

No the child is not correct. Ratio of sides is 4:1. Area of smaller triangle is 10 ft^2. Ratio of areas is $4^2 : 1^2$,

in this case the ratio of areas would be $40^2 : 10^2$ or 1,600 ft^2 to 100 ft^2.

LE6   Skill

a) 1:3

b) 6 sq cm and 54 sq cm

c) 1:9

d) 1 cu cm, 27 cu cm

e) 1:27

LE7   Reasoning

a) 3:2

b) 9:4

c) $m^2 : n^2$

d) 27:8

e) $m^3 : n^3$

LE8   Skill

a) 81:144

b) 27:64

c) $\dfrac{27}{64} = \dfrac{x}{160}$

     $64x = 4,320$

        $x = 67.5 \text{ m}^3$

LE9   Connection

Length of body, length of nose.

LE10   Opener

For student

LE11   Connection

   a) $A = \pi r^2$

   b) $A = \pi(1)^2$
      $= \pi \text{ in}^2$

   c) $A = \pi(2)^2$
      $= 4\pi \text{ in}^2$

   d) The adult has a muscle cross section that is 4 times larger in area than that of the young dolphin's, so the adult is 4 times stronger.

LE12   Concept

   a) 1:4

   b) 2

LE13   Connection

   a) For student

   b) Volume relationships

   c) For student

LE14   Reasoning

   32

LE15   Reasoning

   27

LE16   Connection

   a) 10,000

   b) 1,000,000

   c) A cockroaches weight has increased 100 times more than its strength.

LE17   Summary

   Ratios of similar solids can provide useful information about the solids relationships.

Homework Exercises 10.7

1.   a) $\dfrac{3}{5} = \dfrac{12}{x}$

            $3x = 60$

              $x = 20$

     b) $1:4$

     c) $1:16$

     d) $1:4$

3.   a) $16\pi : 36\pi \rightarrow 4\pi : 9\pi \rightarrow 4:9$

     b) $2:3$

     c) $9 \div 4 = 2.25$ times more

5. $A_1 = \pi(4.5)^2$

          ≈ 63.6 in^2

Costs \$3.49 or \$0.054 per square inch

     $A_2 = \pi(6)^2$

          ≈ 113 in^2

Costs \$5.79 or \$0.051 per square inch

     $A_3 = \pi(8)^2$

          ≈ 201 in^2

Costs \$7.99 or \$0.039 per square inch

     a) Pizza with 16" diameter is best buy.

     b) Pizza with 9" diameter is worse buy.

7.   a) 3

     b) 9

9.   a) $3:4$

b) $A_1 = 2(9)(6) + 2(6)(3) + 2(3)(9)$

$= 198 \text{ m}^2$

$A_2 = 2(12)(8) + 2(8)(4) + 2(4)(12)$

$= 352 \text{ m}^2$

c) $\dfrac{198}{352} = \dfrac{9}{16}, \; 9:16$

d) Square of the ratio of their edges.

e) $V_1 = 9(6)(3)$

$= 162 \text{ m}^3$

$V_2 = 12(8)(4)$

$= 364 \text{ m}^3$

f) $\dfrac{162}{384} = \dfrac{27}{64}$

g) Cube of the ratio of the edges.

11.   a) $4:1$

b) $64:1$

13.   a) $\dfrac{5}{8}$

b) 218.75

c) Deduction

15. $\dfrac{h_2}{h_1} = 3$ Then $SA_2 = 3(3)SA_1$

$= 3(3)(40)$

Surface area of new prism is 360 ft^2.

$V_2 = 3(3)(3)V_1$

$= 3(3)(3)(24)$

Volume of new prism is 648 ft^3.

17.   (b) Areas

19.   a) $\dfrac{5}{4} = 1\dfrac{1}{4}$ Larger

       b) $\dfrac{25}{16} = 1\dfrac{9}{16}$ Stronger

       c) $\dfrac{125}{64} = 1\dfrac{61}{64}$ Heavier

       d) Smaller one

21.   a) 3

       b) 9

       c) 27

23.   Gulliver's volume is $12^3 = 1{,}728$ times the volume of the Lilliputian.

25.   a) Volume

       b) Ratio of diameters: $\dfrac{26}{20} = \dfrac{13}{10}$

       Ratio of volumes: $\dfrac{2{,}197}{1000} = 2.197$

       If child's bicycle is \$60, then adult bicycle should be $2.2 \times \$60 = \$131.82$

27.   a) Choice (3) volume.

       b) Choice (2) area.

29.   Large

31.   For student

Chapter 10 Review Exercises

1.   a) 0.237 m

       b) 800,000 mg

3.   b

5. Enclose the triangle with a rectangle. Find the area of the rectangle and subtract the areas of the three "external" triangles from the rectangle's area.

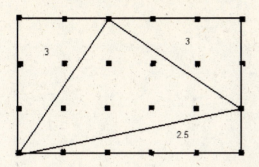

$15 - (3 + 3 + 2.5) = 6.5$ units^2

7. A is the better buy.

    Package A:

    1 roll is 14 sq ft

    4 rolls are 56 sq ft

    56 sq ft costs $5

    1 sq ft costs $0.089

    Package B:

    1 roll is 14 sq ft

    3 rolls are 42 sq ft

    42 sq ft costs $4

    1 sq ft costs $0.095

9. Put together two of the triangles to form a parallelogram.

    $$A_\triangle = \frac{1}{2} A_{parallelogram}$$
    $$= \frac{1}{2} bh$$

11. $A_{square} = (10)(10)$

          $= 100 \ m^2$

    $A_{circle} = \pi r^2$

          $= \pi(5)^2$

          $= 25\pi \ m^2$

    $A_{shade} = (100 - 25\pi) \ m^2$

13. $50 = 2\pi r$

    $r = \dfrac{25}{\pi} \ cm$

    $A = \pi r^2$

      $= \pi\left(\dfrac{25}{\pi}\right)^2$

      $= \dfrac{625}{\pi} \ cm^2$

15. a) In each triangle, $m\angle 1 + m\angle 2 = 90^0$ so $\angle DAB$ is a right angle.

    b) Area of $ABCD = c^2$

    c) $EFGH + \triangle ADH, \triangle ABE, \triangle BCF, \triangle CDG$

    d) $b - a$.

    e) $4\left(\dfrac{1}{2}ab\right) + (b-a)^2$

    f) $c^2 = 4\left(\dfrac{1}{2}ab\right) + (b-a)^2$

       $c^2 = 2ab + b^2 - 2ab + a^2$

       $c^2 = b^2 + a^2$

17. $A_{circle} = \pi(15)^2$

          $= 225\pi \ units^2$

    $A_{triangle} = \dfrac{1}{2}(18)(24)$

           $= 216 \ units^2$

    $A_{shade} = (225\pi - 216) \ units^2$

19. $A = 12 + 8 + 4 + 4 + 8 + 4 + 16 + 16$

      $= 72 \ units^2$

21. $$166 = 2(4)(5) + 2(4w) + 2(5w)$$
       $$126 \text{ m}^2 = 18w$$
       $$7 \text{ m} = w$$

23. $V = bh$ applies to prisms and cylinders.

25. $$V = bh$$
       $$= \pi r^2 h$$
       $$= \pi(36)(12)$$
       $$= 432\pi \text{ in}^3$$

       $$A = 2\pi(36) + 12\pi(12)$$
       $$= 72\pi + 144\pi$$
       $$= 216\pi \text{ in}^2$$

27.   a) $\dfrac{200}{675}$

       b) $\dfrac{400}{900} = \dfrac{4}{9}$

       $\dfrac{\sqrt{4}}{\sqrt{9}} = \dfrac{2}{3}$

       $\left(\dfrac{2}{3}\right)^3 = \dfrac{8}{27}$

       $\dfrac{27}{8} \times 200 = 675 \text{ m}^3$

Chapter 11

Lesson Exercises 11.1

LE 1   Opener

Your age is a quantity that is variable because it changes. The number of fingers you have is a quantity that

is constant because it doesn't change.

LE2   Concept

a) A is a variable.

b) π is a constant.

c) r is a variable.

LE3   Concept

a) y represents an unknown.

b) The variables represent a general property.

c) A formula that relates quantities.

d) Quantities that vary in relation to one another.

LE4   Skill

a) Let D represent the daily allowance of iron, $0.25D$

b) $0.25(20) = 5$

c) Iron is not a number. A variable must represent a quantity.

LE5   Concept

Not possible.

LE6   Skill

$P = \$230 + 0.14s$

LE7   Skill

S = sum of money spent by the U.S. government on transportation and education in 2009.

N = amount spent on national defense.

$$S < \frac{1}{6}N$$

LE8   Concept

How the money is spent within each department; how much of the money goes to large campaign

contributions.

LE9   Skill

$a, b, c, d \in Q$

$b, d \neq 0$

$$\frac{a}{b} \times \frac{c}{d} = \frac{ac}{bd}$$

LE10   Skill

a) y is 16 more than 4 times d.

b) A dog's age in human years is 16 more than 4 times the dog's age in years.

LE11   Connection

a) $C = 2,200 + 2x$

b) $R = 18x$

c) Profit = Revenue − Cost

d) $P = 16x - 2,200$

e) Cell B3 = 2(A3) + 2,200

Cell C3 = 18(A3)

Cell D3 = C3 − B3

g) For student

LE12   Communication

(c)

LE13   Communication

a) Between the 1st and 2nd hour.

b) Between the 2nd and 3rd hour.

c) The car started fast, stopped for an hour and then sped up again.

d) $\dfrac{\text{total distance}}{\text{total time}} = \dfrac{90}{3}$

$\qquad\qquad\qquad = 30$ mph

LE14   Summary

Algebra is a mathematical language. Algebraic thinking involves three processes: 1) doing and understanding; 2) building rules to represent functions; and 3) abstracting from computation. Algebra is limited, it only describes relationships between quantities involved in the given algebraic expression.

Homework Exercises 11.1

1.   a) Constant

      b) Variable

3.   a) Variable

      b) Constant

      c) Variable

5.   a) General property of numbers.

      b) Quantities that vary in relation to one another.

      c) A formula that relates quantities.

      d) Variable is used to represent an unknown number.

7.   A variable must represent a quantity.

9.   a) Let A represent average temperature.

      $A + 12^{o}$

      b) $60^{o} + 12^{o} = 72^{o}$

11.   Let T be total cost, A be adult tickets, and c be children tickets: $T = \$5A + \$3c$

13.   Let P be the perimeter and L the length: $P = 3L$

15.   Let x be last year's height and y be this year's height: $y \geq 2x$

17.   a) x is the number of dishes needed: $8 + x = 13$

       b) x is the number of CDs sold: $\$15x = 90$

19.   a) A is weight after 5 weeks: $A = 200 - w$

       b) Possible answer: $w = 10$

21.   a) T is total ticket cost: $6c + 5(6) = T$

       $$6(c + 5) = T$$

       b) Distributive property

SSM Chapter 11   416

23. a, b, and c are real numbers: $a(b+c) = ab + ac$

25.   No, "is" translates to $=$, $n = 4x$

27.   a) The rate equals 10 plus $22 timed D. R is the rate and D is the number of days.

       b) The rental rate is $10 plus $22 times the number of days.

29.   a) C is greater than 2 times P.

       b) The current U.S. population is greater than twice the U.S. population in 1940.

31.   The difference is that the y value for $5y$ represents a variable the other two values are numbers. $5y$ is the

       product of 5 and some input for y.

33.   a) The car started fast, slowed down for a while and then went fast again.

       b) $\dfrac{120 \text{ miles}}{3 \text{ hours}} = 40 \text{ mph}$

35.   a) 3

       b) 2

       c) 1

37.   a) $50(3) = 150$ miles

       b) 150 sq. units

39. S is the new salary, x is the old salary: $S = 1.5x$

41.   a) $T = \dfrac{300}{5} + 1$

       b) Set up spreadsheet.

       c) $B3 = \dfrac{300}{A3} + 1$

       d) The total time decreases as the speed increases.

43.   a) For student

b) For student

c) (1) Product on top: xy

(2) Product on bottom: $(7-x)(7-y) = 49 - 7x - 7y + xy$

(3) $x(7-y) = 7x - xy$

(4) $y(7-x) = 7y - xy$

Sum: $xy + 49 - 7x - 7y + xy + 7x - xy + 7y - xy = 49$

Lesson Exercises 11.2

LE1   Opener

a) Addition property of equality.

b) Multiplication property of equality.

LE2   Connection

$$2 + 3x = 14$$
$$3x = 12$$
$$x = 4$$

LE3   Reasoning

a) The child adds 2 and 4.

b) The child adds all three numbers.

c) An equal sign represents a relation between numbers, not a signal to carry out computation.

LE4   Reasoning

$$P = 230 + 0.14S$$
$$400 = 230 + 0.14S$$
$$\$1214.29 = S$$

LE5   Concept

a) Yes

b) Yes

LE6   Skill

a) $z - 3 > 6$
   $z > 9$

b)

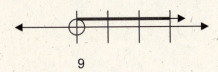

9

LE7   Concept

a) Yes

b) No

c) Yes

d) No

LE8   Skill

a) $-4r \leq 36$
   $r \geq -9$

b)

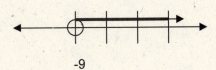

-9

LE9   Summary

Equations and inequalities observe the same solution strategies but we must be careful to understand that

when an inequality is multiplied or divided by a negative value the sign of the inequality must be

accommodated.

Homework Exercises 11.2

1. $x = 1$

3. $16 = 10 + 3x$

    $6 = 3x$

    $2 = x$

    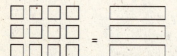

    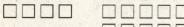

5. a) 12

    b) $12 = 4x$

    $3 = x$

    c)

    $\square + 2 = 8$

    $\square = 6$

    $-3x = 6$

    $x = -2$

7. To isolate x undo multiplication by 2. Divide both sides of the equation by 2.

9. a) W is weekly salary, s is sales: $W = 100 + 0.05s$

    b) $250 = 100 + 0.05s$

    $150 = 0.05s$

    $\$3,000 = s$

11. a) 100 miles is \$32, 400 miles is \$128, 500 miles \$160, so you could drive about 500 miles.

    b) T is total cost and m is miles: $T = 120 + 0.32m$

    $280 > 120 + 0.32m$

    $160 > 0.32m$

    $500 > m$

13. x is number students: $0.36x = 261$
$$x = 725$$

15.   a) $T = \dfrac{N}{4} + 40$

        b) $4(T - 40) = N$

        c) $T = \dfrac{100}{4} + 40$
$$T = 65^0$$

        d) $4(95 - 40) = N$
$$220 = N$$

17.   You could tell the student to add the inverse of the quantity to each side of the inequality and then look at the inequality sign.

19.   a) $-4n \le 20$
$$n \ge -5$$

        b)

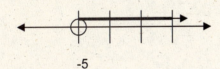

-5

21.   a) $-2t + 30 > 18$
$$-2t > -12$$
$$t < 6$$

        b)

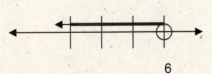

6

23. x represents the number of visits:

        $85 \ge 4x$
     $21.3 \ge x$

     When the number of visits is less than 21 it is cheaper to pay \$4 per visit.

25. $6.1x - 0.23(6.1x) = 100$
$$4.697x = 100$$
$$x \approx 21.3 \text{ hours}$$

27.   a) For student

b) $w = 1 + 2020 + \dfrac{2019}{4} - \dfrac{2019}{100} + \dfrac{2019}{400}$

$w = 1 + 2020 + 504 - 20 + 5$

$w = 2510$

$$7 \overline{)\,2510\,} \quad 358R4$$

Fourth day is Wednesday.

Lesson Exercises 11.3

LE1   Opener

a) A

b) Measure the angle created between the ramp and its base.

c) The slope can be measured by comparing the rise (vertical distance) to the run (horizontal distance).

d) The perpendicular sides are closer to each other in length.

e) They are equal: $\dfrac{2}{3} = \dfrac{4}{6}$

LE2   Concept

a) i) < 0

ii) > 0

iii) > 0

b) Answer follows the exercise

LE3   Reasoning

a) $\dfrac{8-2}{4-1} = \dfrac{6}{3} = 2$

b) $\dfrac{y_2 - y_1}{x_2 - x_1} = \text{slope}$

LE4   Concept

a) It would make the fraction undefined (dividing by 0).

b) A vertical line passes through these points.

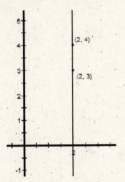

LE5   Reasoning

a)

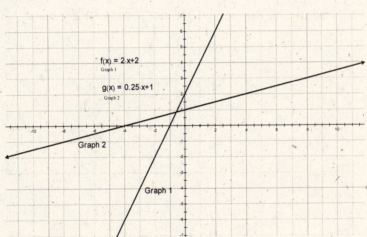

b) The slopes are 2 and one-fourth respectively

c) The slope is represented as *m*, the coefficient of the *x* term.

LE6   Connection

a) 0

b) $(0, b)$

LE7   Skill

a) The slope is -2 and the *y*-intercept is (0, 3).

b) Rise is -2 and the run is 1

c)

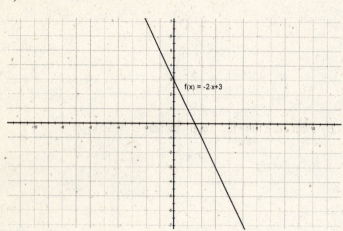

$f(x) = -2 \cdot x + 3$

LE8   Reasoning

a)

w	0	10	20	30
g	40	39	38	37

b) $g = 40 - \dfrac{w}{10}$

c) Slope: $-\dfrac{1}{10}$; g intercept (0, 40)

d) The ground clearance decreases by 1 cm each time the weight increases by 10 kg.

e)

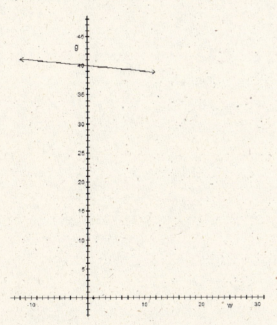

LE9   Reasoning

a) (1) Linear

(2) Nonlinear

(3) Linear

b) You can recognize a linear pattern if y changes by a fixed amount when x changes by a fixed amount.

LE10   Summary

Slope measures the steepness of a line. The slope is represented by ratio, $\dfrac{rise}{run}$. $\dfrac{y_2 - y_1}{x_2 - x_1} = m$ represents the

slope of a line joining (x_1, y_1) and (x_2, y_2). $y = mx + b$ represents the slope-intercept form of a line. b is the

y-intercept, m is the slope. Parallel lines have the same slope.

Homework Exercises 11.3

1.   a) $\dfrac{6}{12} = \dfrac{1}{2}$

      b) $\dfrac{6}{12} = \dfrac{x}{30}$

           $12x = 180$

               $x = 15$ feet

3. $\dfrac{3}{x} = \dfrac{1}{12}$

      $x = 36$ feet

      $c^2 = 3^2 + 36^2$

      $c^2 = 1,305$

      $c \approx 36.12$ feet

5.   a) > 0

      b) < 0

      c) $= 0$

7.   a) $m = \dfrac{2}{3}$

      b) $m = -\dfrac{4}{3}$

      c) $m = \dfrac{2}{3}$

9.   a)

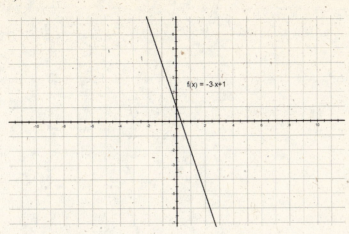

f(x) = -3·x+1

   b) The slope is -3.

   c) The shortcut is to look at the *x* coefficient.

11.   a) The slope is one-half and the *y* intercept is (0, 2).

   b) To graph the line, first plot the *y* intercept then move up one-half unit and to the right one unit.

13. $y = \dfrac{2}{3}x$

15.   a) (4, 0)

   b) Yes

   c) An infinite number.

17.   a)

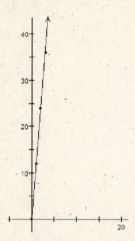

b) $m = \dfrac{24-12}{2-1}$

   $m = 12$

c) How fast he runs.

d) Time cannot be negative.

19.   a) $D = \dfrac{T}{5}$

b)

c) Each change of 1 second is an additional $\dfrac{1}{5}$ mile of distance.

d) The lightning (light) reaches one in virtually no time, so the formula simply estimates the distance the

thunder (sound) travels at a rate of 0.2 miles/second.

21.   a) Yes

b) No

c) Yes

23.   a) Yes

b) The slope between any two points of an arithmetic sequence will be the same. If the slope is the same

then the line is straight.

25. $m = \dfrac{49 - 43}{44 - 42}$

$m = 3$

$w = 3H + b$

$43 = 3(42) + b$

$-83 = b$

$w = 3H - 83$

27. a) Pool 3

b)

Pool Number (x)	1	2	3	4
Number of Tiles (y)	8	12	16	20

c)

Total tiles is equal to 4 times the pool number plus 4

d) $m = \dfrac{12-8}{2-1}$

$m = 4$

$y = 4x + b$

$8 = 4 + b$

$4 = b$

$y = 4x + 4$

e) Induction

f) $y = 4(12) + 4$

$y = 52$

29.   a) -1 and 1: ½ and -2

b) The product of the slopes is -1.

31.   a) $\dfrac{m}{1}$

b) $\dfrac{1}{-m}$

33.   Rotation about the origin

35.   a) – b) For student

c) They are equal.

Lesson Exercise 11.4

LE1   Opener

    a) Increase by 2

    b) Decrease by 3

    c) y does not change a constant amount

LE2   Reasoning

    a)

x	-2	-1	0	1	2
y	2	-1	-2	-1	2

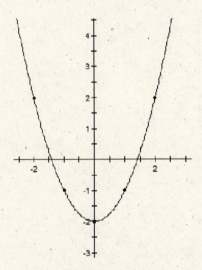

    b) It is shifted down 2 units on the y axis.

    c) $y = x^2 + c$ is the curve $y = x^2$ translated c units vertically.

LE3   Skill

a)

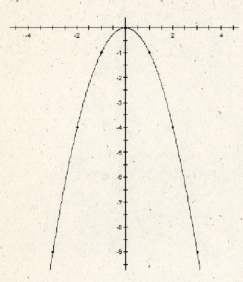

b) $y = x^2$ opens up, $y = -x^2$ opens down

LE4   Reasoning

a) $A = L^2$

b) 3

LE5   Reasoning

a)

x	-1	0	1	2	3
y	$\frac{1}{3}$	1	3	9	27

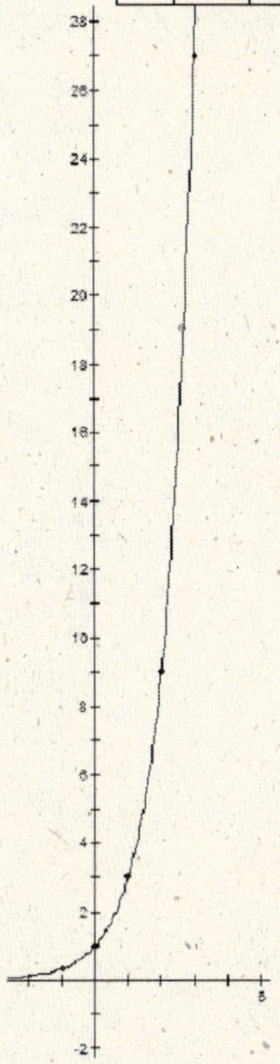

b) They both increase sharply for $x > 0$, get closer to zero for negative values of x. The graph never goes

below the x axis.

c) Both $y = 3^x$ and $y = 2^x$ approach the x axis but never reach it.

d) In the graph $y = a^x$, as x increases by the same amount each time, y will multiply by the same amount

each time.

LE6   Reasoning

a)

t	0	40	80	120
P	6	12	24	489

b) Yes

c)

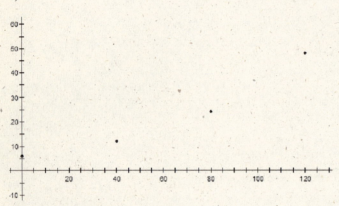

d) $P = 6(2)^{(16/40)}$

   $= 7.92$ billion

e) In 2028

LE7   Connection

a) P – has no particular growth pattern, just is increasing.

Q – the population increases by 2 thousand every 10 years

R – the population triples every 10 years

b) P – neither

Q – linear

R – exponential

LE8   Summary

Linear functions have a constant rate of change. Linear functions have the form $y = mx + b$. The equation

of a quadratic function has the form $y = ax^2 + bx + c$ where $a \neq 0$. The graph of a quadratic function is a

parabola. Exponential functions have to the form $y = a^x$. They model population growth, compound

interest and radioactive decay.

1.　　a)

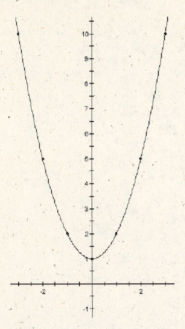

　　b) It is the graph $y = x^2$ translated up 1 unit.

　　c) For $0 < x \leq 3$

3.　　It is the graph of $y = x^2 + 1$ translated 4 units.

5.　　(c)

7.   a) $h = -16(3)^2 + 80(3)$

        $h = -144 + 240$

        $h = 96$ feet

    b)

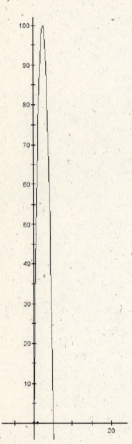

    c) $t = 0.7$ sec

       $t = 4.3$ sec

9.   a)

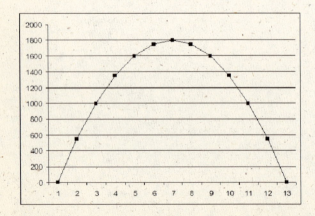

    b) $6

11.   Both of these functions approach the x axis but never reach it.

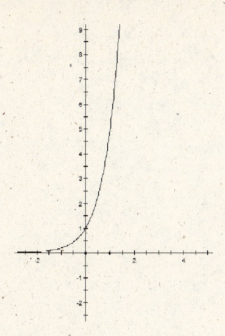

13.   a)

t	0	1	2	3
moths	50	100	200	400

        b) 3

        c) $m = 50(2^{1.5})$
             $= 141$

15.   a) $A = 3,000(1.06)^{20}$
             $\approx \$9,621.41$

        b) $4,000 = 3,000(1.06)^t$

                $\dfrac{4}{3} = 1.06^t$

             $\ln(1.3) = t \ln(1.06)$

             $\dfrac{\ln(1.3)}{\ln(1.06)} = t$

             $4.5 \text{ years} \approx t$

17.   P – exponential

        Q – neither

        R – linear

19.   a)

L	3	5	7	9	11
W	10	6	30/7	10/3	30/11

b) $LW = 30$

$$W = \frac{30}{L}$$

c)

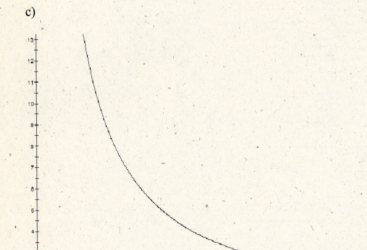

d) The graph is all in the first quadrant, and y decreases less and less rapidly (from left to right). The curve

gets very close to the x axis.

e) Any positive number.

21. a) $WT = 3,000$

$$T = \frac{3,000}{W}$$

b) $T = \dfrac{3,000}{600}$

$T = 5$ min

c)

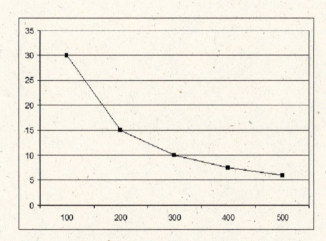

23. a) – c) The results should all be the same.

d) Part (c)

Chapter 11 Review Exercises

1. a) $|x - y| = |y - x|$

b) $|x + y| > x + y$ works for $x = 2, y = -3$ but not for $x = -2, y = 3$.

3. a) $200 \div 4 = 50$ mph

b) No

5. First we should understand where the child got the 7 from, if it is the sum of the 4 and 3, then maybe the distributive law is being misunderstood.

7. x is amount spent on meal.

$$x + 0.25x = 50$$
$$1.25x = 50$$
$$x = \$40$$

$40 to spend on meal.

9.   a) $-3t + 2 > -19$
              $t < 7$

       b)

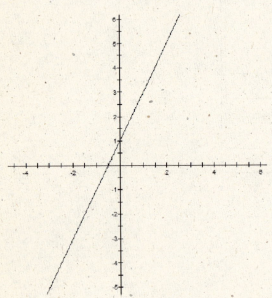

                    7

11.   a)

       b) $y = -2x + 1$

13.   a) For each increase of 1 inch in height, there is a 5.5 pound increase in weight.

       b) $H = \dfrac{W + 220}{5.5}$

       c) The original.

15.

No *x* intercept.

17.   P is neither

       Q is exponential

       R is linear

Lesson Exercises 12.1

LE1   Opener

       For student

LE2   Connection

       For student

LE3   Skill

       For student

LE4   Skill

       For student

LE5   Concept

       Yes, each chose 5 random papers from the bag.

LE6   Reasoning

       a) Yes

       b) The whole class.

LE7   Opener

       a) Write each of the 100 people's names on a piece of paper and draw 20 out of a paper bag. You could select a sample from the day students and a sample from the evening students in proportion to the sizes of the two groups.

       b) Random sampling is the best.

LE8　Skill

$$640 + 416 + 544 = 1600$$

$$\frac{640}{1600} = 40\%$$

$$\frac{416}{1600} = 26\%$$

$$\frac{544}{1600} = 34\%$$

40% of 200 = 80

26% of 200 = 52

34% of 200 = 68

$$80 + 52 + 68 = 200$$

LE9　Reasoning

a) TV station viewers do not accurately represent all adults.

b) The sample is not representative of the entire campus.

LE10　Connection

No. In a random sample each individual has an equally likely chance of being selected. Discuss the

definition of random sample.

LE11　Reasoning

Telephoning people would not be representative in that some adults don't have telephones. Some people

with telephones have unlisted numbers.

LE12　Concept

a) Should people be allowed to smoke cigarettes in restaurants?

b) Should people be permitted to smoke cigarettes in restaurants in spite of the threat to the health of

others?

c) Don't people have the right to smoke cigarettes in restaurants?

LE13　Reasoning

For student

LE14   Opener

         For student

LE15   Concept

         Answer follows the exercise

LE16   Concept

         a) Observational Study

         b) Experiment

LE17 Summary

In an observational study investigators do not apply treatments while observing the variable of interest, whereas in experimental studies investigators apply a treatment then observe the effects of the treatment. Factors that could contribute to bias might include attitude of the researcher, the sample population chosen, or the methods used to collect data.

Homework Exercises 12.1

1.   a) Registered voters in Colorado.

         b) $\dfrac{322}{1200} = 27\%$

3.   a) For student

         b) $20 + 42 + 28 = 90; \dfrac{20}{90} = 22\%$; 22% of 500 = 111 prefer whole milk.

         $\dfrac{42}{90} = 47\%$; 47% of 500 = 233 prefers low fat milk.

         $\dfrac{28}{90} = 31\%$; 31% of 500 = 156 prefers skim milk.

5. $250 + 220 + 200 + 180 = 850$

$\dfrac{250}{850} = 29\%$ Freshman and 29% of 100 = 29 freshman

$\dfrac{220}{850} = 26\%$ Sophomore and 26% of 100 = 26 sophomores

$\dfrac{200}{850} = 24\%$ Junior and 24% of 100 = 24 juniors

$\dfrac{180}{850} = 21\%$ Seniors and 21% of 100 = 21 seniors

7.   a) Simple random sample

b) Stratified sampling

9.   a) Stratified random sample

b) Random sampling

11.   a) Draw 100 names from the list of names of all students at the college.

b) The percent of the student body that are freshman, sophomores, juniors, and seniors are all representative in terms of the 100 chosen.

13.   b

15.   It appears U.S. adults want all industries except for the supermarket industry to be regulated more.

17.   a) Under coverage

19.   a) Not everyone has a telephone, people who have an opinion will likely share others may hang up the phone

b) One hundred is not enough to make a generalization relative to the population of the state

21.   a) The results of a poll can influence what people think and consequently how they respond to subsequent polls.

b) For student

23.   a) Yes

b) No

c) The respondents were not at all representative of the voters.

25.   No, detention is not based on gender but on behavior. Young boys tend to misbehave more often than young girls.

27.   It gives the impression of tennis as a favorite sport.

29.   a) Should aluminum cans be recycled?

        b) Should we improve the clutter in our environment by recycling cans?

        c) Should we waste time by collecting and recycling cans?

31.   Do you have to speak the language fluently?

33.   For student

35.   a) Experiment

        b) Observational

37.   We don't know anything about participant's lifestyles

39.   For student

41.   a) By showing that the 35% who do not respond are similar to the 65% who do.

        b) As many shows as there are TV sets.

43.   None of these. A share is the televisions in use from the Nielson sample, not all TV's in use.

45.   The most likely kind of death that results from keeping a firearm at home is suicide. The next most likely result is killing someone you know in a fight. These events are much more likely than using a firearm to defend your against a burglar.

47.   For student

49.   For student

Lesson Exercise 12.2

LE 1   Opener

        a) Bar graphs, line plots, circle graphs, pictographs

        b) Newspapers, magazines, reports

LE 2   Reasoning

        a) The first line plot shows the response of each child in the class with regard to their family size. The second line plot shows the relative frequency of different family signs.

        b) The first graph.

        c) The second graph.

LE 3   Skill

a)

```
 0| 8
 1| 249
 2| 47
 3| 2
 4| 269
 5| 5
 6| 26
 7| 9
 8| 9
 9|
10| 0
11| 2
12|        6
13|
14|
15|
16| 6
17|
18|
19| 8
```

b) There are more incomes on the lower end of the scale

LE 4   Concept

   Answer follows exercise

Annual Income (thousands of dollars)	Frequency
1 – 35	5
36 – 70	6
71 – 105	3
106 – 140	2
141 – 175	1
176 – 210	1

LE 6   Skill

a) $153 \div 9 = 17$

Annual Income (thousands of dollars)	Frequency
7 – 24	5
25 – 42	4
43 – 60	4
61 – 78	1
79 – 96	1
97 – 114	2
115 – 132	1
133 – 150	0
151 – 168	2

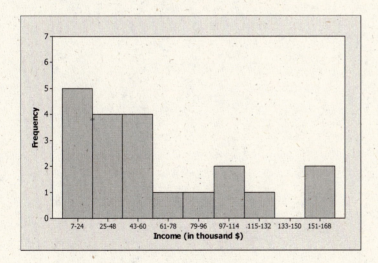

b) It is more uniform, but it does have a gap in the 133-150 interval.

LE 7   Concept

Graphs (b) and (c) are histograms.

LE8   Skill

There are 16 adult members of the church in this age range.

LE 9   Skill

a) Most likely there would be a person born every month of the year.

b) Answers will vary.

c) Answers will vary.

d) Answers will vary.

LE 10   Skill

a)

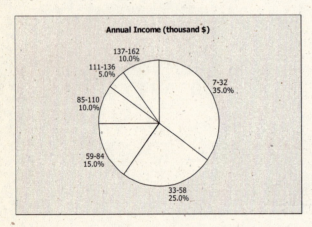

b) Answers will vary.

LE11   Reasoning

a) What was the population in 1990 ?

b) Estimate the population in 1994

c) Predict the population in 2010

d) The population has been increasing from 1980 to 2000.

LE 12   Opener

Graph (b) is more appropriate since the bar graph is used when more than one activity is involved.

LE 13   Concept

1. A circle graph

2. A line graph

3. A bar graph

a)

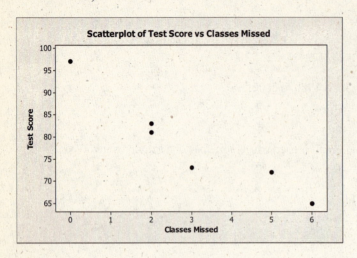

b)

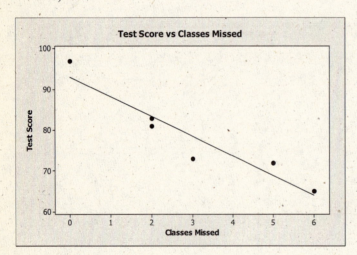

c) Negative slope

d) Decreases

e) Using (2, 83) and (0, 97) gives a slope of $m = \dfrac{97-83}{0-2} = \dfrac{14}{-2} = -7$.

The slope is –7.

$$y = -7x + b$$
$$97 = -7(0) + b$$
$$97 = b$$
$$\Rightarrow y = -7x + 97$$

f) As more classes are missed, test scores drop.

g) $y = -7(4) + 97$

$\quad y = 69$

The test score is 69.

h) It is too far beyond the data range.

i) Answers will vary.

LE 15   Concept

a) Positive

b) Negative

LE 16   Summary

Bar graphs are used to compare the value of several variables.

A graph is a simple visual summary of a set of data. It should have a title and label on both axes.

Circle graphs are useful for showing what part of the whole falls into each of several categories.

Line graphs are most often used to show changes over a period of time.

12.2 Homework Exercises

1. a) Education

   b) Mechanical Engineering

   c) The salaries range from $22,505 to $34,460, the highest starting salary being about 55% higher than the lowest. Technological fields tend to pay higher salaries.

   d) Answers will vary.

   e) Answers will vary.

   f) Answers will vary.

3. a) 2

   b) 3

   c) 1

5. a)

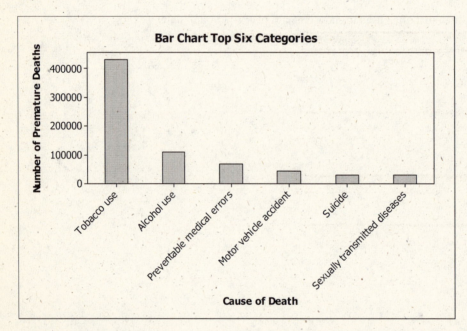

   b)   Answers will vary.

   c) Tobacco use is the major cause of premature deaths. Alcohol contributes to more deaths than illegal drug use. Suicide accounts for more deaths than murder.

   d) It would be useful to know how the numbers were derived, and the breakdown of types of murders.

7.   a) Class A

  5|8
  6|22
  7|025
  8|015
  9|28

     b) Class B

   4|2
  5|
   6|558
   7|5
   8|02
  9|019

     c) Class B has more extreme scores than Class A.

9.   Graph (c) is the only graph that is a histogram.

11.   a)

Interval	Frequency
60 – 69	5
70 – 79	1
80 – 89	2
90 – 99	4

     b)

Interval	Frequency
60 – 79	6
80 – 99	6

13.   29% of 360 = 104.4°

15. a)

Food	25%
Clothing	8.3%
Rent	50%
Entertainment	6.7%
Other	10%

b)

	Central angles
Food	90
Clothing	30
Rent	180
Entertainment	24
Other	36

17. For Student

19. a) The middle 20% stayed about the same.

The upper 5% increases in value somewhat.

The upper 1% earned a lot more money.

b) For student, perhaps tax breaks for the wealthiest.

21. For student

23. a) Circle graph

b) Line graph

c) Bar graph

25. a) Bar; Stem & Leaf

   b) Bar, Stem & Leaf, Line

   c) Circle

27. a)

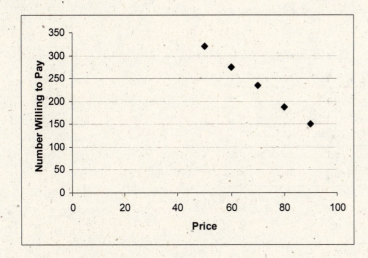

   b)

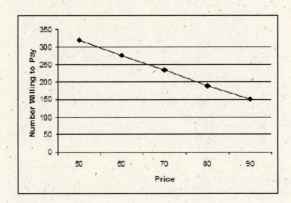

   c) Negative

   d) Decrease

   e) Negative

   f) $m = \dfrac{180 - 320}{80 - 50}$

   $m = \dfrac{-14}{3}$

   $y = \dfrac{-14}{3}x + 553$

   g) The higher the price, the less the demand.

SSM Chapter 12   457

h) Demand = 296

i) Too far out of range.

j) For student

29.   a)

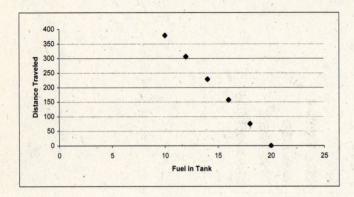

b)

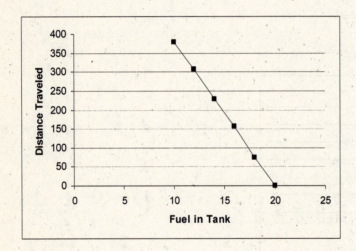

c) Negative

d) $m = \dfrac{379 - 0}{10 - 20}$
   $m = -37.9$

e) For every gallon the fuel decreases the distance increases 38 miles.

f) $379 + 38 = 417$ miles traveled when 9 gallons are left.

g) For student

31.   a) Positive

        b) Negative

33.   a) E, T, A, O

        b)

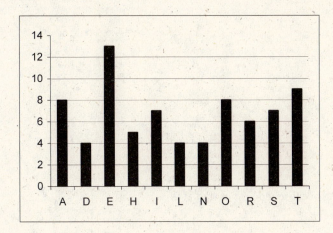

        c)

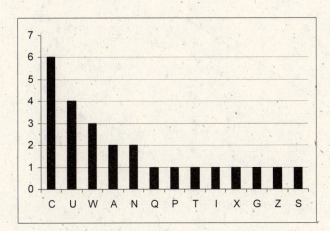

        d) C

        e) The meeting will be at the pier.

35.   a) mean for x is 6, mean for y is 78.5

        b)

37.   For student

39.   a) For student

        b) 2010 – 299 million

        2020 – 323 million

        $$y = \left(4.92 \times 10^{-7}\right)(1.01)^x$$

Lesson Exercises 12.3

LE 1   Opener

        a) Misleading; The graph is misleading because it does not start at 0.

        b)

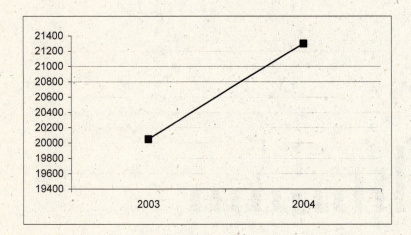

        c) The average pay increase over a period of several years would be helpful to know.

LE2   Concept

a)

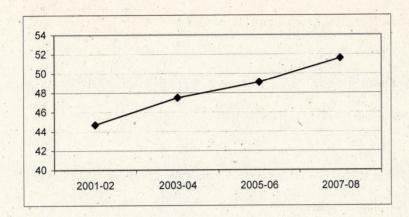

b)

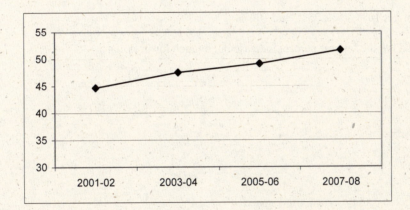

LE3   Connection

Tell the child to indicate the break, this alerts the reader to read the numbers on the axes.

LE4   Reasoning

a) A very generous offer.

b) Assume the pants cost $100 last year. Now they would cost $150. 50% off 150 is $75, and

$150 - 75 = \$75$. Overall there is a 25% decrease from last years price.

LE5   Skill

Suppose food prices were $100, 10% increase is $110. A second 10% increase is $11. Over 2 years the food price is now $121. This represents a 21% increase.

LE6   Concept

No, 4% of the United States GDP is not the same as 4% of Belgium's GDP. Taking the same percent of different numbers does not give the same answer.

LE7   Concept

This is not justified as 25% of your rent is much higher than 25% of your heating bill. The heating bill is only a small part of total expenses.

LE8   Reasoning

a) Last column has 10%, 40%, 25%, and 100%.

b) Suds 'n Crud has the highest percent increase but a much lower dollar.

LE9   Reasoning

a) 8 pounds

b) 0 pounds

LE10   Reasoning

Price, repair record, safety record, trunk space

LE11   Reasoning

The rate of inflation during each time period; whether anything unusual happened between 1976 and 1984.

LE12   Summary

Misleading statistics could lead to poor decision making.

Homework Exercises 12.3

1. a) Israel

   b) Start the vertical axis at 0. The bars will be much closer in height.

   c) U.S.

   d) Length of school day.

3. $\begin{array}{ccc} 100 & 120 & 120 \\ \times 0.20 & \times 0.30 & +36 \\ \hline 20.00 & 36.00 & 156 \end{array}$ Over the two years, oil prices rose 56%.

5. $\begin{array}{ccc} 500 & 500 & 400 \\ \times 0.20 & -100 & \times 0.20 \\ \hline 100.00 & 400 & 80.00 \end{array}$ Her salary will be $480.

7. A percent is a ratio, so it would be more appropriate to average these numbers.

9. $\begin{array}{cc} 0.10 & 0.10 \\ \times 0.60 & -0.06 \\ \hline 0.0600 & 0.04 \end{array}$

   4% profit this year

11. How far each car traveled.

13. a) 50%

    b) 0%

15. a) AIM with fluoride and AIM without fluoride.

    b) Fluoride is effective in reducing dental decay.

17. They were using different time periods.

19. a) 1993 – 2000

    b) 1990 – 1993 or 2000 – 2003

    c) It increased steadily from 1990 to 1993, then began declining from 1993 to 2000. It increased again from 2000 to 2004 then began to decline again.

21. The first region is 4 times bigger in area.

23. $P \to 1.20P \to 1.3(1.2)P = 1.56P$ Which is 56% more than P.

25. For student

Lesson Exercises 12.4

LE1   Opener

   a) For student

   b) Add all the travel times and divide by the number of students. See what travel time was the most

   frequent, the mode.

LE2   Skill

   a) For student

   b) Mean

LE3   Connection

   a) Mode

   b) Median

LE4   Connection

   a) Pick the most common size stack.

   b) Arrange the stacks in size order and pick the middle (third) one.

   c) 4

   d) Mean

LE5   Concept

   Yes, 2 is 2 below 4 and 6 is 2 above 4 which balances out

LE6   Skill

   a) Yes. 3 and 5 balance, 2 and 8 balance

   b) No. 4 and 7 do not balance

   c) Yes. 1, 2, and 4 are a total of 8 below 5 and 8 and 10 are a total of 8 above 5.

LE7   Reasoning

   a) Rocky Ride: Mean = 52,000; Median = 56,000

   Road Rubber: Mean = 55,000; Median = 50,000

   b) Pick Road Rubber if you think all the tires should be counted equally, (mean).

   Pick Rocky Ride if you think the outlier (99,000) is suspicious and you will use the median to lessen its

   impact on the comparison.

LE8   Reasoning

        a) (1) Median (2) Mean

        b) (1) Mean (2) Median

LE9   Connection

        a) The mean $338.89

        b) The median $200

LE10   Reasoning

        For student

LE11   Summary

        Three useful measures of the center are the mean, median, and mode. The mean or average of a set of

        numbers is their sum divided by how many numbers there are, $\bar{x} = \dfrac{s_1 + x_2 + ... + x_N}{N}$. The median is the

        middle value of a set of numbers when the numbers are arranged in order. The mode is the number that

        occurs most frequently.

Homework Exercises 12.4

1.   No, 5 occurs the most frequently

3.   Mode

5.   No, the numbers are not in order.

7.   a) Yes, 60, 60, 80, 80

        b) Yes, 70, 70, 70, 70

        c) No

        d) Yes, 40, 80, 80, 80

9.   a) $\dfrac{9+11}{2} = 10$

        b) $\dfrac{20+25}{2} = 22.5$

        c) $\dfrac{36+100}{2} = 68$

11.   {4, 4, 4, 5, 5, 6, 7}

13. Form 3 stacks of 2 counters, 1 stack of 4 counters, and 1 stack of 5 counters.

   a) Ask the student "which stack occurs most frequently?" to get the mode.

   b) When arranging the stacks by the number of counters in each stack, the middle stack would be the median.

   c) To find the median, trade until all stacks have roughly the same number of counters.

15. a) Yes

   b) Yes

   c) No

17. In most large groups, close to 50% of the people score below average.

19. Mean: equal weight to each score

   Median: uses order of scores, middle

   Mode: most frequent

21. If you sum the hours, your choice would be A, but there are more children in A. So the average time for group A is 19 hours and the average time for group B is 19, so in this regard each group watched the "same" amount.

23. Mean = 20, Mode = 20, Median = 20

   Replacing the 18 with 0 yields

   Mean = 16.4, Mode = 20, Median = 20

25.   a) $\dfrac{77(3)+x}{4}=80$

$231+x=4(80)$

$x=89$

b) $y=\dfrac{3(77)+t}{4}$

$y=\dfrac{231+t}{4}$

c)

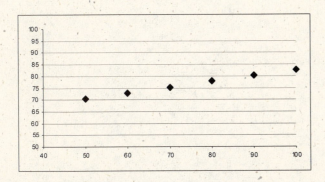

d) If she makes 100 on the 4th test she will have an 82 average. For every 10 points gained on the test her

average increases slightly.

27.   a) 28 senators

b) 3 or $3,000,000

c) 9.46 or 9.46 million

d) 9

e) 6 or $6,000,000

29. $\dfrac{\text{sum boys scores}+\text{sum girls scores}}{28}=73$ Find the difference of 876 and 2044 (the product of 28 and 73) to

$\dfrac{\text{sum girls scores}}{12}=70.5$

sum girls scores=876

get the sum of the boys scores.

sum boys scores=1168

$\dfrac{1168}{16}=73$

The boys average is 73.

31.   85.4 is the minimum score needed on the final.

$$\frac{68+79+88+74+82}{5} = 78.2$$

$$\frac{78.2+78.2+78.2+x}{4} = 80$$

$$234.6 + x = 320$$

$$x = 85.4$$

33. Median

35. Mode

37. a) Median

     b) Median

     c) Mean = 25.6, without outlier = 19.5

     Median = 20, without outlier = 20

     Mean had a bigger change.

39. a) Median

     b) Mean

41. Median

43. a) $\dfrac{10+10}{\dfrac{10}{50}+\dfrac{10}{30}} = \dfrac{20}{\dfrac{1}{5}+\dfrac{1}{3}} = \dfrac{20}{\dfrac{3}{15}+\dfrac{5}{15}} = \dfrac{20}{\dfrac{8}{15}} = 20\times\dfrac{15}{8} = 37.5$ mph

     b) $\dfrac{10+10}{\dfrac{10}{m}+\dfrac{10}{N}}$ or $\dfrac{20mN}{m+N}$

45. $2.8\times45 = 126$ Let x = g.p.a. this semester.

$$\frac{126+15x}{60} = 3$$

$$126+15x = 180$$

$$15x = 54$$

$$x = 3.6$$

He needs a 3.6 g.p.a.

47. a) Class 1: 16, Class2: 11, Class 3: 7.5, Class 4: 7.5

b) Class 1: 14.25, Class2: 14.25, Class 3: 8.5, Class 4: 5

c) Class 4 prefers trickle down plan. Class 2 and 3 prefer drag down plan. Class 1 prefers to leave things as they are.

d) 1

e) 3 and 4

49. For student

51. For student

53. For student

Lesson Exercises 12.5

LE1   Opener

a) Molly: Mean =7, Median =7

Billy: Mean = 7, Median = 7

b) Molly's scores have more variation. Billy is more consistent.

LE2   Concept

a) 60% of children in the test group scored below your child.

b) Above average

LE3   Skill

a)

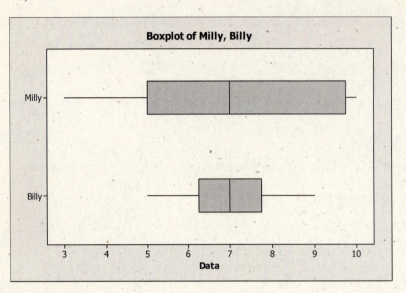

b) Molly's box plot is longer, showing a greater variance from the median. Billy's box plot is more compact, showing greater consistency as his scores are closer to the median.

LE4   Reasoning

The child is taking the mean of all the scores below the median rather than finding the median below the median to get the 1st quartile. The child is taking the mean of all the scores above the median rather than finding the median above the median to get the 3rd quartile.

LE5   Concept

a) Milly 4.5; Billy 1

b) Milly's scores had a larger spread than Billy's

LE6   Skill

No, 3 is only 2 below the first quartile

LE7   Skill

$$3-7=-4, 5-7=-2, 5-7=-2, 6-7=-1,$$
$$8-7=1, 9-7=2, 10-7=3, 10-7=3$$

LE8   Skill

Sum of the deviations is 0.

LE9   Skill

a) Billy's standard deviation would be lower than Molly's

b) Billy's standard deviation

Score	Deviation from Mean	Squared Deviation
5	-2	4
6	-1	1
7	0	0
7	0	0
7	0	0
7	0	0
8	1	1
9	2	4
		10 Sum

$$\text{Mean} = \frac{10}{8} = 1.25$$

$\sqrt{1.25} = 1.118$, Billy's standard deviation is 1.118.

LE10   Concept

a) Teenytown Hamsters $\dfrac{320}{5} = 64$

Leantown Lizards $\dfrac{320}{5} = 64$

b) Teenytown Hamsters heights have a larger standard deviation because there is a greater range in their

heights.

LE11   Opener

a) Most heights would cluster around an average height.

b)

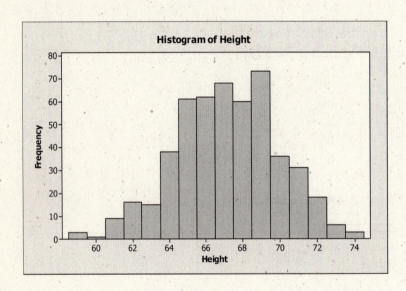

LE12   Skill

$\dfrac{121}{500}$

LE13   Connection

a) About 68% have heights within 1 standard deviation of the mean.

b) 95% have heights within 2 standard deviations of the mean.

LE14   Skill

a)

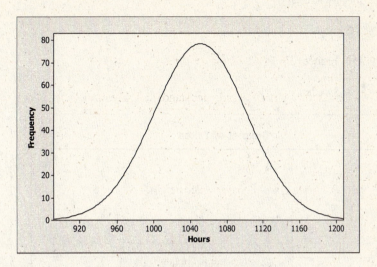

About 95% of the light bulbs will last between 950 and 1,150 hours.

b) 16%

c) Deduction

LE15   Skill

a) 50th

b) 97.5^{th}

LE16   Connection

a) Skewed to the right.

b) Symmetric

c) Skewed to the left.

LE17   Summary

Statisticians usually describe a set of data using a measure from the center (mean or median). Sometimes a more detailed assessment is needed and a 5 – number summary will be used to describe a set of data. The standard deviation is the most common measure statisticians use to describe the spread.

Homework Exercises 12.5

1.   Their results should be comparable

3.   c

5.   Lyle scored better in mathematics than 43% of those in his group.

7.   a) Min 23% US

1st quartile = 30% Switzerland

Median = 78% Germany

3rd quartile = 45% France

Maximum = 51% Sweden

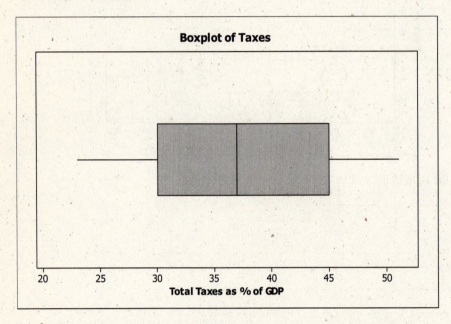

b) The U.S. has the smallest percent of total taxes as its GDP. It ranks slightly below Japan.

c) What services the country provides to its citizens.

9.   a)

	Class A	Class B
Min	56	68
1st Q	70	74
Median	81	80
3rd Q	91	86.5
Max	97	91

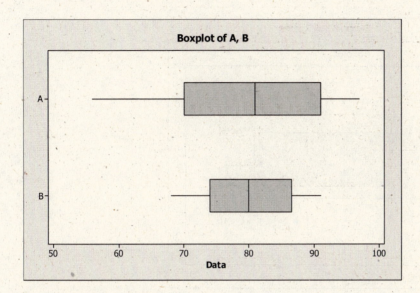

b) Class B is more consistently close to the median in their scores.

11.   a) Min = 64

Q1 = 72

Median = 80

Q3 = 86

Max = 92

b)

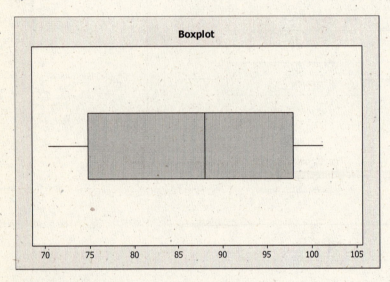

c) The second box and whisker plot is shifted to right and more spread out (10% more).

13.   a)

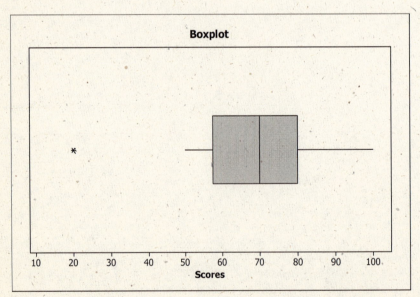

b)   There are is one suspected outlier according to the 1.5(IQR) Rule. The score of 20 lies below Q_1 –

1.5(IQR).

15. a) Box and whisker plot

    b) Box and whisker plot

17. b

19. b has the smallest standard deviation

    c has the largest standard deviation

21. 68% of Brand A light bulbs will last between 980 and 1,020 hours. A is more likely to last close to 1,000 hours. 68% of Brand B light bulbs will last between 900 and 1,200 hours. B lasts longer than 1,000 hours on average.

23. Class with wide ability range, high standard deviation. Honors class, high mean and low standard deviation. Students with similar abilities, low standard deviation. Remedial class, low standard deviation, low median.

25. a) (i) has the larger standard deviation.

Mean (i) = 4

Score	Deviation from Mean	Squared Deviation
1	-3	9
2	-2	4
4	0	0
9	5	25
		38	Sum

$38 \div 4 = 9.5$

$\sqrt{9.5} = 3.082$

Mean (ii) = 4

Score	Deviation from Mean	Squared Deviation
2	-2	4
4	0	0
5	1	1
5	1	1
		6	Sum

$6 \div 4 = 1.5$

$\sqrt{1.5} = 1.225$

b) 3 numbers are within 1 standard deviation of the mean.

c) Range (i) = 9 − 1 = 8

Range (ii) = 5 − 2 = 3

27. Set 1: 1, 1.5, 2, 3, 4

Set 2: 1, 1, 2.5, 3, 4

29.   a) The mean increases by $2,000 and the standard deviation stays the same.

        b) The mean increases by 10% and the standard deviation increases by 10%.

31.   For student

33.   a) 68%

        b) 2.5%

        c) 2.5%

35.   a) 50%

        b) 16%

        c) 2.5%

37.   a) 16th percentile

        b) 97.5 percentile

        c) 48th percentile

39.   a) Skewed to the left

        b) Normal

        c) Skewed to the right.

        d) Skewed to the right.

41.   Skewed to the right.

43.   Skewed to right.

45.   a) 7

        b) 2.125

47. a) $\bar{x} = \dfrac{A+B+C+D}{4}$

b) $s = \sqrt{\dfrac{(A-\bar{x})^2 + (B-\bar{x})^2 + (C-\bar{x})^2 + (D-\bar{x})^2}{4}}$

c)

$$s = \sqrt{\dfrac{A^2 - 2A\bar{x} + (\bar{x})^2 + B^2 - 2B\bar{x} + (\bar{x})^2 + C^2 - 2C\bar{x} + (\bar{x})^2 + D^2 - 2D\bar{x} + (\bar{x})^2}{4}}$$

$$s = \sqrt{\dfrac{A^2 + B^2 + C^2 + D^2 - 2\bar{x}(A+B+C+D) + 4(\bar{x})^2}{4}}$$

$$s = \sqrt{\dfrac{A^2 + B^2 + C^2 + D^2 - 2\bar{x}(4\bar{x}) + 4(\bar{x})^2}{4}}$$

$$s = \sqrt{\dfrac{A^2 + B^2 + C^2 + D^2 - 4(\bar{x})^2}{4}}$$

$$s = \sqrt{\dfrac{A^2 + B^2 + C^2 + D^2}{4} - (\bar{x})^2}$$

47. Adding k to each score increases the mean by k, so the deviation of a score x_i from the still be $x_i - \bar{x}$. If all the

deviations stay the same, the standard deviation will also stay the same.

49.   For student

Lesson Exercises 12.6

LE1   Opener

        For student

LE2   Concept

        a) 35/44

        b) She got 35 right out of 44.

LE3   Concept

        a) 60

        b) Fran's scored higher in math computation than 60% of the students in this age group.

LE4   Reasoning

        No. Their percentile scores are too close to be considered significantly different. Since the tests are

        multiple choice, guessing accounts for part of each child's score.

LE5   Skill

        6

LE6   Skill

        a) 0.1

        b) $88.9 - 60 = 28.9$

LE7   Concept

        a) 5.4

        b) He scored as well as the average student in the fourth month of fifth grade.

LE8   Reasoning

The student is being compared to third grade students based upon a test with very few third grade questions.

LE9   Concept

It is very likely that the students true score is between 390 and 450 (inclusive).

LE10   Connection

a) The stanines are 7, 4, 6, 4, and 6.

b) If your school is average, this student will be fairly average. He may be a little stronger than average in English and social studies and a little below average in science and math.

LE11   Summary

As a teacher you will receive standardized test reports on your students. The raw score is the student's actual score on the exam. A percentile score indicated the percent of students who scored below a given individual. Stanines divide the normally distributed student scores into nine groups based on their percentiles. The grade level score indicates the grade level, in years and months, at which the students score, would be average.

Homework Exercises 12.6

1.   a) 36 out of 60

b) She got 36 questions right out of 60.

3.   Students in this area score below the national average.

5.   a) 7.4

b) She scored the same as the average child in the fourth month of seventh grade.

7.   No

9.   8; 4; 6

11.   No, their confidence bands overlap. Their scores are not significantly different.

13.   a) 5, 5, 4, 4, 6

b) Above average in social studies and a little below average in everything else.

15.   b and c

Chapter 12 Review Exercises

1. Range: $590 - 34 = 556$

    A reasonable number of intervals would be 5.

    $556 \div 5 \approx 111$, Make intervals of length 111.

    Classes:

    34 – 145

    146 – 257

    258 – 369

    370 – 481

    482 – 593

3. a) There was a slight decrease from 1992 to 1994 then a sharp increase for 1996. A downward trend was

    exhibited until 2004 where the trend began to increase again.

    b) The U.S. government spends more money than it collects in taxes, so it borrows money and pays interest on

    it.

5. a) Negative correlation when a worker is sick, production is lowered.

    b) Slope = 3.5. For every sick day productivity drops 3.5 units.

7. a) Suppose sugar prices were $1, a 40% increase would be $0.40.

Then 1.40 would mean a drop of 28 cents.

$$\begin{array}{r} \times 0.20 \\ \hline 0.28 \end{array}$$

The final price is now at $1.40 - 0.28$ or $1.12. This is 12% rise.

b) $P \rightarrow 1.4P \rightarrow 1.12P$; Overall change is 12%

9.

```
14|0
13|02
12|45688
11|58
10|
```

11. $15 \times 9 = 135$ and $135 - 10 = 125$. 125 is the sum of the other 14 numbers.

13. Median

15. He scored better than 59% of students in his category in reading.

17. Mean $= \dfrac{5+6+10}{3} = 7$

Standard deviation

Score	Deviation from Mean	Squared Deviation
5	-2	4
6	-1	1
10	3	9
		14 Sum

$14 \div 3 = 4.667$

$\sqrt{4.667} = 2.16$ or 2.2

19. a) 95%

b) 16%

21. a) \$97,920 is the median and \$113,422 is the mean. There are more house prices far above these numbers than

below them (skewed to the right). These high prices pull up the mean more than the median.

b) Skewed to the left.

23. $850 + 800 + 750 = 2400$

$\dfrac{850}{2400} = 35.4\%$ and 35.4% of 80 = 28 sophomores

$\dfrac{800}{2400} = 33\%$ and 33% of 80 = 27 juniors

$\dfrac{750}{2400} = 31.3\%$ and 31.3% of 80 = 25 seniors

25. a) Do you support an increase in funding for government regulation of food safety?

b) Do you support protecting the health of our citizens by increasing funding for government regulation of food

safety?

c) Do you support an additional tax burden by increasing funding for government regulation of food safety?

27. Grade level scores are misleading in that people expect most children should score at or above grade level even

though about half of the students should score below grade level. Very high or low grade level scores have

little meaning.

SSM Chapter 12   485

Chapter 13

Lesson Exercises 13.1

LE1   Opener

        a) Pulling 1 blue marble

        b) Pulling 1 black marble

        c) Pulling 1 yellow marble

        d) Getting a blue or black marble

LE2   Skill

        a) {HH, HT, TH, TT}

        b)

First Toss	Second Toss	Outcome
Heads	Heads	HH
	Tails	HT
Tails	Heads	TH
	Tails	TT

LE3   Concept

        a) No

        b)

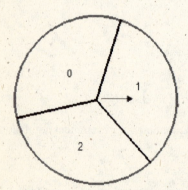

LE4   Concept

$$P(2) = \frac{1}{4}$$

$$P(2) = \frac{1}{3}$$

LE5   Concept

a) $P(H) = \frac{1}{2}$ If you flip 10 times, you should get 5 heads.

b) For student

c) For student

LE6   Concept

a) For student

If the student got 6 heads the $P(H) = \frac{6}{10}$ where 10 was the total number of tosses.

b) For student

LE7   Connection

a) For student

b) Bruce is correct. The probability of an event is the favorable outcomes divided by the total outcomes.

When two coins are flipped the total outcomes are {HH, HT, TH, TT}.

c) The second model where each outcome is equally likely.

LE8   Skill

a) $P(\text{sum of } 7) = \frac{6}{36}$ or $\frac{1}{6}$

b) – d) For student

LE9   Reasoning

a)

	1	2	3	4	5	6
1	2	3	4	5	6	7
2	3	4	5	6	7	8
3	4	5	6	7	8	9
4	5	6	7	8	9	10
5	6	7	8	9	10	11
6	7	8	9	10	11	12

b) 36 outcomes

c) Yes

d) $P(\text{sum of } 7) = \dfrac{6}{36}$ or $\dfrac{1}{6}$

e) For student

LE10   Skill

a)

Sum	2	3	4	5	6	7	8	9	10	11	12
Probability	$\dfrac{1}{36}$	$\dfrac{2}{36}$	$\dfrac{3}{36}$	$\dfrac{4}{36}$	$\dfrac{5}{36}$	$\dfrac{6}{36}$	$\dfrac{5}{36}$	$\dfrac{4}{36}$	$\dfrac{3}{36}$	$\dfrac{2}{36}$	$\dfrac{1}{36}$

b) The answers are symmetric.

c) $P(\text{sum of } 11) = \dfrac{2}{36}$ So, $50 \times \dfrac{2}{36} = \dfrac{100}{36} = 2\dfrac{28}{36}$ or about 3 times.

LE11   Connection

(d)

LE12   Reasoning

She can look at attendance records for the past month of school days and see what percent of the days her teachers were in school.

LE13   Reasoning

a) $P(\text{female}) = \dfrac{94}{1000} = \dfrac{241}{250} = 0.964$

b) $P(\text{math/science}) = \dfrac{226}{1000} = 0.226$

c) $P(\text{female/math/science}) = \dfrac{212}{964} = 0.22$

LE14   Reasoning

a) Not fair, squares a better deal.

b) For student

c) Possible products: 1, 2, 3, 4, 5, 6, 7, 8, 9, 10, 12, 15, 16, 18, 20, 24, 25, 30, 36

Possible squares: 1, 4, 9, 16, 25, 36

$P(\text{product of } 36) = \dfrac{1}{18} = 6\%$

$P(6^2) = \dfrac{1}{6} = 17\%$

The likelihood of a winning square is greater.

LE15   Summary

The probability of an event is the number of favorable outcomes divided by the number of possible outcomes. The theoretical probability of an event approximates the fraction of the time this event is expected to occur when the same experiment is repeated many times uniform conditions.

Homework Exercises 13.1

1.   a) Roll a number less than 4

b) Roll a 6

c) Roll a 7

d) Roll a number less than 7

3.   a) {1, 2, 3, 4, 5, 6}

b) Yes

5. a) Not equally likely

    b) Not equally likely

    c) Equally likely

7. i) $\dfrac{1}{4}$

    ii) $\dfrac{1}{6}$

    iii) $\dfrac{1}{8}$

    iv) $\dfrac{165}{300} = \dfrac{33}{72} = \dfrac{11}{24}$

    b) For student

9. a) $P(\text{between 2 and 9}) = \dfrac{24}{50} = \dfrac{12}{25}$

    b) $P(\text{between 3 and 7}) = \dfrac{12}{50} = \dfrac{6}{25}$

11. $P(\text{Lake}) = \dfrac{4}{18} = \dfrac{2}{9}$

13.

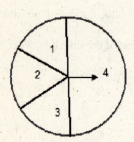

15. The 11 sums are not equally likely.

17. c – both results are equally likely.

19. a) For student

b) For student

c)

First Coin	Second Coin	Third Coin
H	H	H
H	H	T
H	T	H
H	T	T
T	H	T
T	T	H
T	H	H
T	T	T

d) $P(2\text{H}) = \dfrac{3}{8}$

e) For student

21. a) For student

b) For student

23. a)

	1	2	3	4	5	6
1	0	1	2	3	4	5
2	1	0	1	2	3	4
3	2	1	0	1	2	3
4	3	2	1	0	1	2
5	4	3	2	1	0	1
6	5	4	3	2	1	0

b) For student

25. a)

$$P(\text{sum of 2}) = \frac{2}{36} = \frac{1}{18}$$

$$P(\text{sum of 3}) = \frac{4}{36} = \frac{1}{9}$$

$$P(\text{sum of 4}) = \frac{6}{36} = \frac{1}{6}$$

$$P(\text{sum of 5}) = \frac{6}{36} = \frac{1}{6}$$

$$P(\text{sum of 6}) = \frac{6}{36} = \frac{1}{6}$$

$$P(\text{sum of 7}) = \frac{6}{36} = \frac{1}{6}$$

$$P(\text{sum of 8}) = \frac{4}{36} = \frac{1}{9}$$

$$P(\text{sum of 9}) = \frac{2}{36} = \frac{1}{18}$$

b) $\frac{1}{9} \times 100 \approx 11$ times

27. {HHH, HHT, HTH, THH, HTT, TTH, HTT, TTT}

$$P(\text{at least 2H}) = \frac{4}{8} = \frac{1}{2}$$

29. a) {BB, Bb, bB, bb}

b) $P(\text{brown eyes}) = \frac{3}{4}$

c) {BB, Bb, BB, Bb}

$P(\text{brown eyes}) = 1$

31. $P(\text{sum of 4 or 6, or 4, or 6}) = \dfrac{30}{36} = \dfrac{15}{18}$

sum of 4	sum of 6	4	6
1,3	1,5	4,1	6,1
2,2	2,4	1,4	1,6
3,1	3,3	4,2	6,2
	4,2	2,4	2,6
	5,1	4,3	6,3
		3,4	3,6
		4,4	6,4
		4,5	4,6
		5,4	6,5
		4,6	5,6
		6,4	6,6

33.   a) There is 1 head out of 2 possible outcomes.

   b) You will not necessarily get 50 heads. You may not even get close to 50 heads.

   c) If this experiment is repeated a large number of times you will usually get heads about ½ the time.

35.   It rained on 60% of the days that were like this in the past.

37.   a) $\dfrac{300}{500} = \dfrac{3}{5} = 0.60$

   b) $\dfrac{220}{500} = 0.44$

   c) $\dfrac{280}{480} \approx 0.58$

39.   a) 1 5 10 10 5 1

   b) 1, 2, 1

   c) It's the second row.

   d) If you flip a coin 3 times, add up the numbers in the third row to get the denominator (8), and the numerators of the probabilities are the numbers in the third row: 1 for 0 heads, 3 for 1 head, 3 for 2 heads, and 1 for 3 heads.

   e) Inductive

41. One die has 1, 2, 2, 3, 3, and 4. The other has 1, 3, 4, 5, 6, and 8.

43. $P(\text{sum of } 6) = \dfrac{10}{6 \cdot 6 \cdot 6} = \dfrac{10}{216} = \dfrac{5}{108}$

(1, 1, 4)	(2, 1, 3)	(3, 1, 2)	(4, 1, 1)
(1, 2, 3)	(2, 2, 2)	(3, 2, 1)	
(1, 3, 2)	(2, 3, 1)		
(1, 4, 1)			

45. For student

Lesson Exercises 13.2

LE1   Opener

a) Probabilities could be 0% to 100%.

$P(\text{rain on Tuesday}) \; 0 \le P \le 1$

b) Zero, one

LE2   Concept

b, c

LE3   Concept

a) 0

b) Men and women

c) 0

LE4   Concept

a) 1

b) 50%

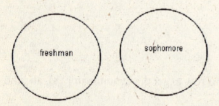

c) No, it cannot be determined since some of the freshman are also women.

LE5   Skill

$$P(1 \text{ or } 2) = P(1) + P(2) = \frac{1}{4} + \frac{1}{2} = \frac{3}{4}$$

LE6   Reasoning

a) $1 - \dfrac{1}{4} = \dfrac{3}{4}$

b) $1 - n$

LE7   Reasoning

a) $P(A) + P(\overline{A})$

b) 1

c) $P(A) + P(\overline{A}) = 1$

LE8   Connection

a) For student

b) H, TH, TTH

c) – d) For student

f) – g) For student

LE9   Connection

a) – d) For student

e) {0, 1, 2, 3, 4, 5}

LE10   Summary

The probability of an event ranges between 0, if it is impossible, to 1 if it is certain. Mutually exclusive events are events that cannot occur at the same time. P (A and B) for mutually exclusive events is P(A) + P(B). The complement of event A is $\overline{A}$, the ways event A does not occur.

Homework Exercises 13.2

1.   (b) , (c)

3.   a) No

        b) Yes

5.   If you add the percentages, you are assuming that no one has black hair and brown eyes which is not true.

7.   0.45

9. $P(female) = \dfrac{1180}{1020 + 1180} = 0.536$

11.   For student

13.   a) Possible events:

        For 1 card -1

        For 2 cards $11, 12, 22 \rightarrow 2 + 1 = 3$

        For 3 cards $111, 112, 113, 122, 123, 133 \rightarrow 3 + 2 + 1 = 6$

        For 6 cards $6 + 5 + 4 + 3 + 2 + 1 = 21$

        $P(6 \text{ different cards}) = \dfrac{1}{21}$, approximately 21 boxes

15.   a) The contestant chose door 1 and door 3.

        b) $\dfrac{3}{5}$

17.   a) – b) For student

19.

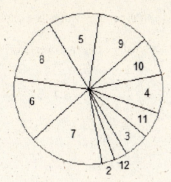

Sum	Degrees
2	10
12	10
3	20
11	20
4	30
10	30
5	40
9	40
6	50
8	50
7	60

21.   a) – d) For student

23.

Sum	Event
2	Triple
3	Double
4	Walk
5	Ground out
6	Single
7	Strikeout
8	Fly out
9	Ground out
10	Fly out
11	Out or double play
12	Homerun

Lesson Exercises 13.3

LE1   Opener

They use the number of letters in the alphabet (26) as a factor the number of times there are slots for letters.

So if there are three slots, 26 is a factor three times or $26 \times 26 \times 26$. The same method is used for the

numbers.

LE2   Connection

a)

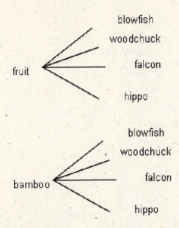

b)

Appetizer	Main dish	Outcome
F	B	FB
F	W	FW
F	F	FF
F	H	FH
B	B	BB
B	W	BW
B	F	BF
B	H	BH

c) Multiply the number of appetizers by the number of entrees.

LE3   Concept

No. There are 5 dessert choices. The child is multiplying the cake choices by the pie choices. You do not

multiply because you are not picking a cake and then a pie in sequence. You are only picking one dessert.

LE4   Skill

$$26 \bullet 26 \bullet 10 \bullet 10 \bullet 10 \bullet 10 = 26^2 \bullet 10^4 = 6,760,000$$

LE5   Skill

$$8 \bullet 10 \bullet 10 \bullet 8 \bullet 10 \bullet 10 \bullet 10 = 8^2 \bullet 10^5$$

LE6   Skill

$$30 \bullet 29 \bullet 28 = 24,360$$

LE7   Skill

a) $_{30}P_3 = 30 \bullet (30-1) \bullet ... \bullet (30-3+1)$
   $_{30}P_3 = 30 \bullet 29 \bullet 28$
   $= 24,360$

b) For student

LE8   Reasoning

a) $2^6 = 64$

b) $\dfrac{6}{64} + \dfrac{1}{64} = \dfrac{7}{64}$

LE9   Skill

a) $5! = 5 \bullet 4 \bullet 3 \bullet 2 \bullet 1 = 120$

b) For student

LE10   Skill

a) 12

b) 24

LE10   Skill

a) $_{30}C_3 = \dfrac{_{30}P_3}{3!}$
   $_{30}C_3 = \dfrac{30 \bullet 29 \bullet 28}{3 \bullet 2 \bullet 1}$
   $= 4,060$

b) For student

LE11   Skill

a) Permutation $_{10}P_3 = 10 \bullet 9 \bullet 8 = 720$

b) Combination $_{100}C_4 = \dfrac{100 \bullet 99 \bullet 98 \bullet 97}{4!}$
$= 3,921,225$

c) Permutation $_5P_5 = 5 \bullet 4 \bullet 3 \bullet 2 \bullet 1 = 120$

LE12   Summary

The fundamental counting principle states we can multiply our choices to find the total number of arrangements. An ordered arrangement of people or objects is called a permutation, for example choosing a president, vice – president, and secretary from a group of 10 people. An arrangement of people or objects where order does not matter is called a combination. It is similar to a permutation except you divide by the number of ways you can arrange the "n" objects or people that are chosen. This is represented by $n! = n(n-1)(n-2) \bullet ... \bullet (2)(1)$.

Homework Exercises 13.3

1. a)

```
                              SK1 ──────── S1SK1
                              SK2 ──────── S1SK2
                    S1 ⟨      SK3 ──────── S1SK3
                              SK4 ──────── S1SK4

        ⟨
                              SK1 ──────── S2SK1
                              SK2 ──────── S2SK2
                    S2 ⟨      SK3 ──────── S2SK3
                              SK4 ──────── S2SK4
```

b)

Shirt	Skirt
1	1
1	2
1	3
1	4
2	1
2	2
2	3
2	4

c) 8 different ways

3. $3 \bullet 4 \bullet 2 = 24$

5. $9 \bullet 9 \bullet 9 \bullet 9 \bullet 9 = 59,049$

7. $_{26}P_4 = 26 \bullet 25 \bullet 24 \bullet 23 = 358,800$

9. $_6P_6 = 6 \bullet 5 \bullet 4 \bullet 3 \bullet 2 \bullet 1 = 720$

11. a) $_nP_n = \dfrac{n!}{n} = \dfrac{n!}{1} = n!$

    b) Deduction

13. a) $8 \bullet 2 \bullet 10 = 160$

    b) Yes

    c) $8 \bullet 10 \bullet 10 = 800$

15. One possible solution:

letter, letter, letter, letter, letter, letter

$26 \cdot 26 \cdot 26 \cdot 26 \cdot 26 = 11,881,376$

17. Use Pascal's triangle

$$\frac{6}{32} = \frac{3}{16}$$

19. a) $5^4 = 625$

b) 3 or 4

c) $P(\text{at least 3 right}) = \frac{17}{625}$

21. $_{52}C_5 = 52 \cdot 51 \cdot 50 \cdot 49 \cdot 48 = 2,598,960$

23. a) When order matters.

b) The number of permutations.

25. $_8C_2 = \dfrac{8!}{2!(8-2)!} = \dfrac{8 \cdot 7}{2!} = 28$

27. $_{12}P_2 = 12 \cdot 11 = 132$

29. a) 1 4 6 4 1

b) $_4C_0 = 1$
$_4C_1 = 4$
$_4C_2 = 6$
$_4C_3 = 4$
$_4C_4 = 1$

c) $_3C_2 = 3$
$_5C_3 = 10$

31.   a) $20 \bullet 20 \bullet 20 = 8,000$

b) $2 \bullet 2 \bullet 1 = 4$

c) $1 \bullet 1 \bullet 1 = 1$

d) $1 \bullet 2 \bullet 2 = 4$

e) $\dfrac{1}{2000}, \dfrac{1}{8000}, \dfrac{1}{2000}$

f) $2 \bullet 18 \bullet 19 + 18 \bullet 2 \bullet 19 + 18 \bullet 18 \bullet 1 = 684 + 684 + 324$
$$= 1692$$

g) $2 \bullet 2 \bullet 19 + 18 \bullet 2 \bullet 1 + 2 \bullet 18 \bullet 1 = 76 + 36 + 36$
$$= 148$$

h) $\dfrac{1692}{8000} = \dfrac{423}{2000} ; \dfrac{148}{800} = \dfrac{37}{200}$

33.   a) 6

b) 1

c) 4

d) 1

e) 2

f) 1

g) $6 \bullet 1 \bullet 4 \bullet 1 \bullet 2 \bullet 1 = 48$

35. $3 \bullet 3 \bullet 3 + 2 \bullet 2 \bullet 2 + 1 \bullet 1 \bullet 1 = 36$

Lesson Exercises 13.4

LE1   Opener

a) Yes

b) No

LE2   Concept

a) Independent

b) Dependent since a 3 on the green die makes it more likely (1/6) that you will get a sum of 9.

LE3   Reasoning

a) Janet

b) Each time the coin is tossed there is a 1 in 2 chance of heads regardless of what happened previously, the probabilities for each toss are the same.

LE4   Reasoning

a) Independent

b) {(1, 1)(1, 2)(1, 3)(2, 1)(2, 2)(2, 3)(3, 1)(3, 2)(3, 3)}

First Draw	Second Draw	Outcome
1	1	1, 1
1	2	1, 2
1	3	1, 3
2	1	2, 1
2	2	2, 2
2	3	2, 3
3	1	3, 1
3	2	3, 2
3	3	3, 3

c) $P(A) = \dfrac{1}{3}$

$P(B) = \dfrac{1}{3}$

$P(A \text{ and } B) = \dfrac{1}{3} \bullet \dfrac{1}{3} = \dfrac{1}{9}$

d) $\dfrac{1}{9}$ of the rectangle represents a second spin of 2.

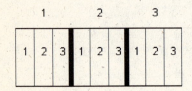

e) $\dfrac{1}{3} \bullet \dfrac{1}{3} = \dfrac{1}{9}$

f) $P(A \text{ and } B) = P(A) \bullet P(B)$

LE5   Skill

a) $P(3, \text{odd}) = \dfrac{1}{6} \bullet \dfrac{3}{6} = \dfrac{1}{12}$

b) $\dfrac{3}{36} = \dfrac{1}{12}$

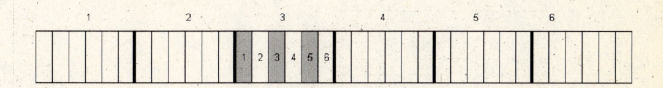

LE6   Reasoning

a) Dependent

b)

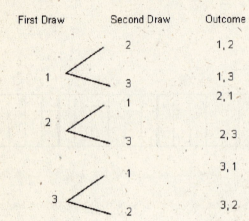

First Draw   Second Draw   Outcome

1
   2   1, 2
   3   1, 3

2
   1   2, 1
   3   2, 3

3
   1   3, 1
   2   3, 2

c) $P(A) = \dfrac{1}{3}$

$P(B \text{ given } A) = \dfrac{1}{2}$

$P(A \text{ and } B) = \dfrac{1}{3} \bullet \dfrac{1}{2} = \dfrac{1}{6}$

d)

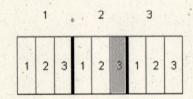

$$\dfrac{1}{3} \bullet \dfrac{1}{2} = \dfrac{1}{6}$$

e) $\dfrac{1}{2}, \dfrac{1}{3}, \dfrac{1}{6}$

f) $P(A \text{ and } B) = P(A) \bullet P(B \text{ given } A)$

LE7   Reasoning

$$\dfrac{6}{16} \bullet \dfrac{5}{15} = \dfrac{3}{8} \bullet \dfrac{1}{3} = \dfrac{1}{8}$$

LE8   Reasoning

a) The independent events formula is a special case of the multiplication rule for probabilities in which

P(B given A) is simplified to P(B).

b) P(A) would have no effect on P(B).

LE9   Reasoning

a) For student

b) For student

c) $\dfrac{6}{10} \cdot \dfrac{5}{9} + \dfrac{4}{10} \cdot \dfrac{3}{9} = \dfrac{30}{90} + \dfrac{12}{90} = \dfrac{42}{90} = \dfrac{7}{15}$

LE10   Summary

Two events are independent if the probability one remains the same regardless of how the other turns out. If

A and B are independent events $P(A \text{ and } B) = P(A) \bullet P(B)$. For dependent events

$P(A \text{ and } B) = P(A) \bullet P(B \text{ given } A)$.

Homework Exercises 13.4

1.   a) Dependent

b) Independent

3.   a) B = it rains tomorrow

b) B = the sum of two regular die is odd

5.   a) Independent

        b) {(1, 1)(1, 2)(2, 1)(2, 2)}

        c)

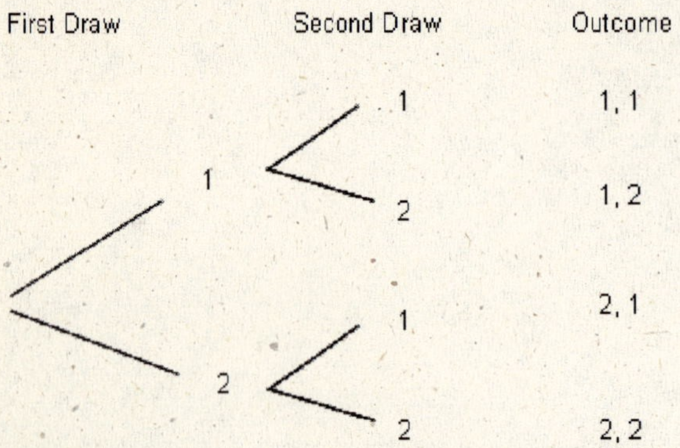

First Draw	Second Draw	Outcome
	1	1, 1
1	2	1, 2
	1	2, 1
2	2	2, 2

        d) $\dfrac{1}{2} \bullet \dfrac{1}{2} = \dfrac{1}{4}$

        e)

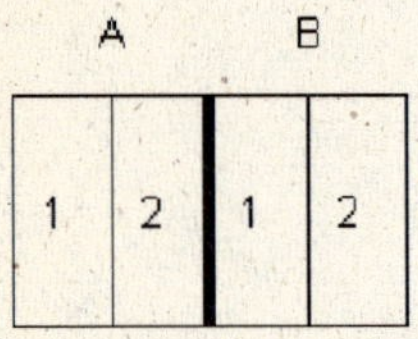

A		B	
1	2	1	2

        f) $P(1 \text{ and } 2) = P(1) \bullet P(2) = \dfrac{1}{2} \bullet \dfrac{1}{2} = \dfrac{1}{4}$

7. $P(\text{both 5 or 6}) = \dfrac{2}{6} \bullet \dfrac{2}{6} = \dfrac{1}{9}$

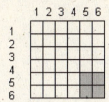

9. a) $0.1 \times 0.7 = 0.07$

   b) $0.9 \times 0.3 = 0.27$

   c) $0.1 \times 0.3 + 0.7 \times 0.7 = 0.03 + 0.63 = 0.66$

11. No, $P(1,1) = \dfrac{1}{6} \bullet \dfrac{1}{6} = \dfrac{1}{36}$. The problem deals with two "consecutive" ones.

13. a) Dependent

    b) $\{(1, 2)(1, 3)(1, 4)(2, 1)(2, 3)(2, 4)(3, 1)(3, 2)(3, 4)\}$

    c) $P(A) = \dfrac{1}{4}$

    $P(B \text{ given } A) = \dfrac{1}{3}$

    $P(A \text{ and } B) = \dfrac{1}{4} \bullet \dfrac{1}{3} = \dfrac{1}{12}$

    d)

1			2			3			4		
2	3	4	1	3	4	1	2	4	1	2	3

    e) $P(A \text{ and } B) = \dfrac{1}{4} \bullet \dfrac{1}{3} = \dfrac{1}{12}$

15. $P(A, A) = \dfrac{4}{52} \bullet \dfrac{3}{51} = \dfrac{12}{2652} = \dfrac{1}{22}$

17. a) Independent

    b) Dependent

19.   a) $\dfrac{5}{13} \bullet \dfrac{4}{12} = \dfrac{5}{39}$

       b) $\dfrac{5}{39} + \dfrac{8}{13} \bullet \dfrac{7}{12} = \dfrac{19}{39}$

21. $P(accident) = 0.01 \bullet 0.006 \bullet 0.002 \bullet 0.002 = 2.4 \times 10^{-10}$

23. $P(\text{mother and daughter attended college}) = P(A) \bullet P(B \text{ given } A)$
$$= 0.3 \times 0.7$$
$$= 0.21$$

25.   a) $(0.7)^5 = 0.168$

       b) $P(at\ least\ a\ 4) = P(4) + P(5)$
$$= 5(0.7^4 \times 0.3) + 0.7^5$$
$$= 0.53$$

27. $\dfrac{97}{100} \bullet \dfrac{97}{100} \bullet \dfrac{97}{100} \bullet \dfrac{97}{100} \bullet \dfrac{97}{100} = \dfrac{8587340257}{10,000,000,000} = 0.86$

29.   a) $P(0) = 40\%$

       b) $P(\text{2 free throws}) = \dfrac{60}{100} \times \dfrac{60}{100} = \dfrac{3600}{10000} = 36\%$

       c) $P(\text{1 free throw}) = 60\%$

31.   You need 3 bulbs.

    $P(\text{1 flowering bulb with 1 bulb}) = 0.60$

    $P(\text{1 flowering bulb with 2 bulbs}) = 0.60 \times 0.4 + 0.4 \times 0.6 + 0.6 \times 0.6$
$$= 0.24 + 0.25 + 0.36$$
$$= 0.84$$

    $P(\text{1 flowering bulb with 3 bulbs}) = 0.4 \times 0.6 \times 0.6 + 0.6 \times 0.4 \times 0.6 + 0.6 \times 0.6 \times 0.4 +$
$$0.4 \times 0.4 \times 0.6 + 0.4 \times 0.6 \times 0.4 + 0.6 \times 0.6 \times 0.6$$
$$= 0.144 + 0.144 + 0.144 + 0.096 + 0.096 + 0.216$$
$$= 0.936$$

33.   a) $P(\text{2 games}) = 0.36$

       b) $P(\text{3 games}) = 0.144 + 0.144$
$$= 0.288$$

35.   a) B = Mike will gain 5 pounds next semester

b) B = Mike will go to school next semester

c) Impossible

Lesson Exercises 13.5

LE1   Opener

a) Not fair

$$P(\text{same color}) = \frac{2}{4} \times \frac{1}{3} + \frac{2}{4} \times \frac{1}{3} = \frac{4}{12} = \frac{1}{3}$$

$$P(\text{opp colors}) = \frac{2}{4} \times \frac{2}{3} + \frac{2}{4} \times \frac{2}{3} = \frac{8}{12} = \frac{2}{3}$$

Player 2 has the advantage

b) For student

c) For student

d) Sample space: $\{(B_1, B_2)(B_1, R_1)(B_1, R_2)(R_1, B_1)(R_1, B_2)(R_1, R_2)\}$

$$P(\text{same}) = \frac{2}{6} = \frac{1}{3}$$

$$P(\text{opp colors}) = \frac{2}{3}$$

LE2   Reasoning

Player 1 wins if an even is rolled.

Player 2 wins if an odd is rolled.

LE3   Concept

a) $3 (win once)

b) Answer follows exercise

c) Answer follows exercise

LE4   Concept

a) $9 payoff for 6 games; $1.50 per game

b) Probabilities are both $\frac{1}{6}$

c) $2 - $1.50 = $0.50 loss per game

LE5   Reasoning

a) $P(winning) = \dfrac{1}{5} \bullet \dfrac{1}{5} = \dfrac{1}{25}$

b) $E = \dfrac{1}{25}(10)$

c) $\$1.00 - \$0.40 = \$0.60$

LE6   Reasoning

$E = \dfrac{1}{6}(\$3) + \dfrac{1}{6}(\$6) + \dfrac{4}{6}(\$0)$

$E = \dfrac{3}{6} + 1$

$E = \$1.50$

Charge $1.50 + 0.50 = \$2.00$

LE7   Concept

a) If the odds are accurate Wet Blanket can be expected to lose 4 times for every 1 time he wins under these conditions.

b) 1 to 4

LE8   Skill

$P(winning) = \dfrac{1}{8}, 0.125, 12.5\%$

LE9   Skill

Odds against winning $\dfrac{37}{1}$

LE10   Summary

In a fair game between two players neither player has the advantage. The expected value of a game is a measure of how much you could expect to win or lose if the game were played many times.

Homework Exercises 13.5

1. $P(divisible\ by\ 3\ or\ 5) = \dfrac{19}{36}$ This game favors you as the probability for a sum divisible by 3 or 5 is greater than ½.

3.   a) Neither player has the advantage.

   b) $P(R,R) = \dfrac{3}{4} \cdot \dfrac{2}{3} = \dfrac{1}{2}$

   $P(B,R) = \dfrac{1}{4} \cdot 1 = \dfrac{1}{4}$

   $P(R,B) = \dfrac{3}{4} \cdot \dfrac{1}{3} = \dfrac{1}{4}$

   $P(\text{same color}) = \dfrac{1}{2}$

   $P(\text{different color}) = \dfrac{1}{4} + \dfrac{1}{4} = \dfrac{1}{2}$

5. $E = \dfrac{1}{10}(1000) + \dfrac{1}{10}(500)$

   $E = \dfrac{1500}{10}$

   $E = \$150$

7.   a) $E = \dfrac{1}{2}(2) + \dfrac{1}{4}(3) + \dfrac{1}{4}(4)$

   $E = 2.75$

   b) $E = \dfrac{1}{3}(1) + \dfrac{1}{3}(3) + \dfrac{1}{3}(4)$

   $E = 2.66$

   c) $E = \dfrac{1}{2}(2) + \dfrac{1}{6}(1) + \dfrac{1}{6}(3) + \dfrac{1}{6}(5)$

   $E = 2.50$

   A is the best spinner to spin.

9.   a) $E = \dfrac{1}{100}(500)$

         $E = \$5$

         $\$25 - \$5 = \$20$ profit

     b) $E = \dfrac{1}{1000}(30,000)$

         $E = \$30$

     c) Part (b) protects you from a possible catastrophic loss, while part (a) does not.

11. $E = 0(0.21) + 1(0.32) + 2(0.18) + 3(0.11) + 4(0.11) + 5(0.07)$

        $E = 1.80$

13. $E = \dfrac{1}{2}(5)$

        $E = \$2.50$

15.   a) $10 \bullet 10 \bullet 10 = 1,000$

     b) $\dfrac{1}{1,000}$

     c) $E = \dfrac{1}{1,000}(500)$

         $E = \$0.50$

     d) $1.00 - 0.50 = \$0.50$

17.   a) $P(black) = \dfrac{18}{38} = \dfrac{9}{19}$

     b) $E = \dfrac{9}{19}(2) + \dfrac{20}{38}(0)$

         $E = \dfrac{18}{19}$

         $E = 0.95$

        $1.00 - 0.95 = \$0.05$ loss

     c) $E = \dfrac{69}{38}(6)$

         $E = \dfrac{36}{38}$

         $E = 0.95$

        $1.00 - 0.95 = \$0.05$ loss

19. $E = \dfrac{3}{36}(10) + \dfrac{6}{36}(7)$

$E = \dfrac{72}{36}$

$E = \$2$

21. $E = \dfrac{4}{52}(10) + \dfrac{12}{52}(5)$

$E = \dfrac{100}{52}$

$E = \$1.92$

$\$1.92 +).40 = \$2.32$ charge per game

23. 1 out of every 12

25. a) 3:2

   b) 2:3

27. a) 9.950

   $\times 0.02$

   199

   199 will test a false positive

   b) $\dfrac{50}{199}$

   c)   70

   $\times 0.02$

   1.40

   d) $\dfrac{29.4}{30.8} \approx 0.95$

29. People choosing surgery have a better chance of surviving more than 3 years.

31. About $52,000

33. For student

Chapter 13 Review

1. {AB, BA, AC, CA, AD, DA, BC, CB, BD, DB, CD, DC}

3. $\dfrac{4}{16}$ or $\dfrac{1}{4}$

5. a) $\dfrac{60}{145} = \dfrac{12}{29}$

    b) $\dfrac{60}{300} = \dfrac{1}{5}$

7. $P(\text{neither chicken nor turkey}) = \dfrac{11}{20}$

9. Average $\dfrac{5+7+4+5+8}{5} = \dfrac{29}{5} \approx 6$ boxes

11. $26 \bullet 26 \bullet 10 \bullet 10 + 26 \bullet 26 \bullet 10 \bullet 10 \bullet 10 = 67,600 + 676,000 = 743,600$

13. $_{20}C_{12} = \dfrac{20 \bullet 19 \bullet 18 \bullet 17 \bullet 16 \bullet 15 \bullet 14 \bullet 13 \bullet 12!}{12! \bullet 8!}$
    $= 125,970$

15. Dependent, independent

17.

Math Class	Pizza Parlor	Pizza Parlor	Pizza Parlor	Rock Concert

    a) $P(\text{math class}) = \dfrac{1}{5}$

    b) $P(\text{pizza parlor}) = \dfrac{3}{5}$

    c) $P(\text{rock concert}) = \dfrac{1}{5}$

19. $0.1 \times 0.1 = 0.01$

21. a) 0.8

    b) $0.2 \times 0.1 = 0.02$

23. $E = \dfrac{1}{200}(500) + \dfrac{1}{1000}(1000)$
    $E = \$3.50$

    Charge $\$10 + \$3.50 = \$13.50$

25.   a) 8

b) $\dfrac{10}{11}$

c) 1:6